HANDBOOK OF NATURE STUDY:

EARTH AND SKY

COMPLETE YOUR COLLECTION TODAY!

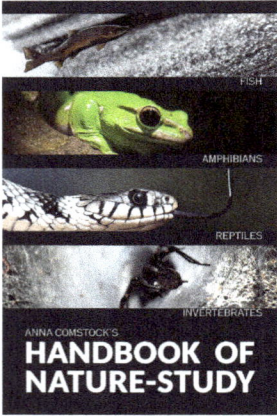

Reptiles, Amphibians, Fish and Invertebrates

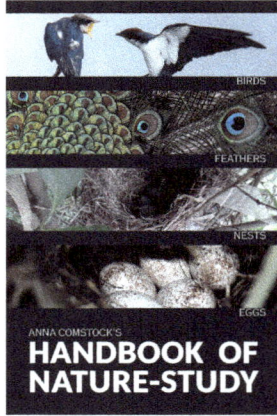

Birds

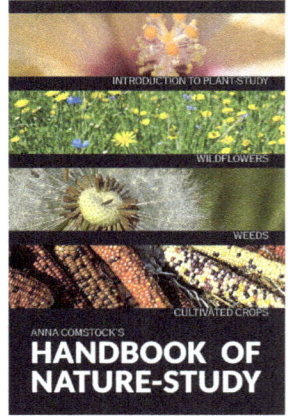

Wildflowers, Weeds and Cultivated Crops

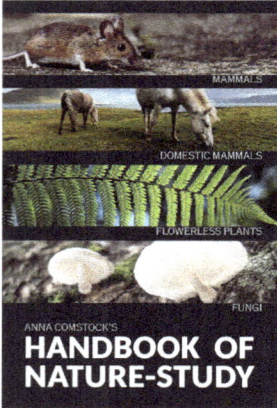

Mammals and Flowerless Plants

Trees and Garden Flowers

Earth and Sky

Insects

Introduction

Available at all online book retailers

LIVING BOOK **PRESS**

OR FROM
LIVINGBOOKPRESS.COM

Handbook of
Nature-Study:
Earth and Sky

———

ANNA BOTSFORD COMSTOCK, B.S., L.H.D

LATE PROFESSOR OF NATURE-STUDY IN CORNELL UNIVERSITY

LIVING BOOK
PRESS

CONTENTS

EARTH AND SKY

The Brook

TEACHER'S STORY

"Little brook, sing a song of a leaf that sailed along,
Down the golden braided center of your current swift and strong."
—J. W. RILEY.

A BROOK is undoubtedly the most fascinating bit of geography which the child encounters; and yet how few children who happily play in the brook—wading, making dams, drawing out the crayfish by his own grip from his lurking place under the log, or watching schools of tiny minnows—ever dream that they are dealing with real geography. The geography lesson on the brook should not be given for the purpose of making work out of play, but to conserve all of the natural interest in the brook, and add to it by revealing other and more interesting facts concerning it. A child who thus studies the brook will master some of the fundamental facts of physical geography, so that ever after he will know and understand all streams, whether they are brooks or rivers. An interesting time to study a brook is after a rain; and May or October give attractive surroundings for the study. However, the work should be continued now and then during the entire year, for each season gives it some new features of interest.

Each brook has its own history, which can be revealed only to the eyes of those that follow it from its beginning to where it empties its water into a larger stream or pond. At its source the brook usually is a small stream with narrow banks; not until it receives water from surrounding hills does it gain enough power to cut its bed deeper into the earth, thus making its banks higher. Where it flows with swift current down a hillside, it cuts its bed deeper, because swift-moving water has more power for cutting and carrying away the soil. However, if the hillside happens to be in the woods, the roots of trees or bushes will help to keep the soil from being washed away. Unless there are obstacles, the course of the brook is likely to be more direct in flowing down a hillside than when crossing level fields. The delightful way in which brooks meander crookedly across the level areas is due to the inequalities of the surface, which interfere more with water on a plain than on a hillside, since the gravity which pulls it forever down has less chance to act upon it forcibly in these situations. After a stream has thus started its crooked course, in time of flood the current strikes with more force against the curves, and cutting them deeper, makes the course still more crooked. The places on the banks where the soil is bare and exposed to the force of the current, are the points where the banks are cut most deeply at flood time.

But the brook is not simply an object to look at and admire; it is a very busy worker, its chief labor being that of a digger and carrier. When it is not carrying anything—that is, when its waters are perfectly clear—the stream is doing the least work. The poets, as well as common people, speak of the playing of the brook when its limpid waters catch the sunbeams on their dimpling surface; but when the waters are roily the brook is working very hard. This usually occurs after a rain, which adds much more water to the volume of the brook; the action of gravity upon this larger and heavier body forces it to flow more swiftly and every drop in the stream that touches the bank or bottom, snatches up a tiny load of earth and carries it along. And every drop thus laden, when it strikes against a corner of the bank, tears more soil loose through the impact, and other drops snatch it up and

Exploring in the brook

carry it on down the stream. And after a time there are so many drops carrying loads and bumping along, knocking loose more earth, that the whole brook, which is made up of drops, looks muddy. In its work as a digger, every drop of water that touches the soil at the bottom or on the banks of the brook uses its own little load of earth or gravel as a crowbar or pickaxe to pry up other bits of dirt and gravel; and all of the drops hastening on, working hard together, dig the channel of the brook wider and deeper. In some steep places, so many of the drops are working together that they are able to pick up pebbles or stones, with which they batter and tear down larger pieces of the bank and scrape out greater holes in the bottom of the stream. On and on the brook flows, a gang of workmen each of which is using its own load as a tool, all in close procession and working double quick. But as soon as the brook reaches a plain or level, its activity ceases; the drops act tired and seem to have no ambition to pick up more soil, and each lets fall its own load as soon as possible, dropping the larger pieces of gravel and rock first, carrying the finer soil farther, but finally letting that down also. If we examine the sediment of a flooded brook, we

find that the gravel is always dropped first, and that the fine mud is carried farthest before it is deposited.

The roar of a flooded stream is very different from the murmur of its waters when they are low. It is not to be wondered at, when we once think of all that is going on in the brook during periods of flood. There are some simple experiments to show what the force of water can do when turned against the soil. Pour water from a pitcher into a bed of soft soil, and note how quickly a hole will be made; if the pitcher is held near the soil, less of a hole will be formed than if the pitcher is held high up, which shows that the farther the water falls, the greater is its force. This explains why the banks of streams are undermined when a strong current is driven against them. The swift current, of course, breaks away more earth at bends and curves than when it is flowing in a straight line; for ordinarily, when flowing straight, the current is swiftest in the bed of the stream, and is therefore only digging at the bottom; but when it flows around curves, it is directed against the banks, and therefore has much more surface to work upon. Thus it is that bends are cut deeper and deeper. If the bare arm is thrust into a flooded brook, we find that many pieces of gravel strike against it; and if we reach the bottom, we can feel the pebbles being moved along over the brook bed.

LESSON

Leading thought— The water from the little brook near our school-house is flowing toward the ocean, and is meanwhile digging out and carrying along with it the soil through which it flows.

Method— The best time to study a brook is after a rain, and October or May is an interesting time for beginning this lesson. The work should be continued during the entire year. It may be done at noon or recess, if the brook is near at hand; or there may be excursions after school, if the brook is at some distance. The observations should be made by the class as a whole.

Observations—

1. Does the brook have its source in a spring or a swamp, or does it receive its water as drainage from surrounding hills? Follow it back to its very beginning. Do you find this in open fields or woods? Is the land about it level or hilly?

2. Are its banks deeper at the beginning, or is the brook at first almost on a level with the surrounding fields? Do the banks become deeper farther from the source? Are the banks higher where the brook flows down hill, or where it is on a level?

3. Is the course of the brook more crooked on a hillside or when flowing through a level area? Are the banks more worn away and steep where the brook flows through woods or bushes than through the open fields?

4. Can you find the places where the water is cutting the banks most, when the brook is flooded? Why does it cut the banks at these particular points?

5. Into what stream, pond or lake does the brook flow? If you should launch a toy boat upon the waters of this brook, and it should keep afloat, through what streams would it pass to reach the ocean? Through what townships, counties, states or countries would it pass?

6. When is the brook working and when is it playing? What is the difference between the color of the water ordinarily and when the brook is flooded? What causes this difference?

7. Make the following experiment to show what the brook is carrying after a storm when the water is roily. Dip from the swift portion of the stream a glass fruit jar full of water. Place it on a window-sill

and do not disturb it until the water is clear. How much sediment has settled at the bottom of the jar? Where was this sediment when you dipped up the water? If this quart of water could carry so much soil or sediment, how much, do you think, would the whole brook carry?

8. Where did the brook get the soil to make the water roily? Study its banks in order to answer this question. Do you think the soil in the water came from the banks that are covered by vegetation or from those which are bare?

9. How did the brook pick up the soil that it carried when it was flooded? Do you think that one of the tools that the brook digs with is the current? Try to find a place where the swift current strikes the bank, and note if the latter is being worn away.

10. Does the swift current take more soil where it is flowing straight, or where there are sharp bends? How are the bends in the brook or creek made?

11. Thrust your bare hand or arm into the swift current of the brook when it is flooded. Do you feel the gravel strike against your arm or hand? Wade in the water. Do you feel the pebbles strike against the feet or legs, as they are being rolled along the bed of the stream?

12. Does the water, loaded with soil and pebbles, dig into the banks more vigorously than just the water alone could do? Which washes away more earth and carries it down stream—a fast or a slow current?

13. Does the water of the brook flow fastest when its waters are low or high? When the brook is at its highest flood, do you think it is working the hardest? If so, explain why. When it is working the hardest and carrying most soil and gravel, does it make a different sound than when it is flowing slowly and its waters are clear?

14. How does the brook look when it is doing the least amount of work possible?

15. Make a map of your brook showing every pool, indicating the places where the current is swiftest and showing the bends in its course. To test the rapidity of the current, put something afloat on it and measure how far it will go in a minute.

16. How many kinds of trees, bushes and plants grow along the banks of your brook? How many kinds of fish and insects do you find living in it? How many kinds of birds do you see frequently near it?

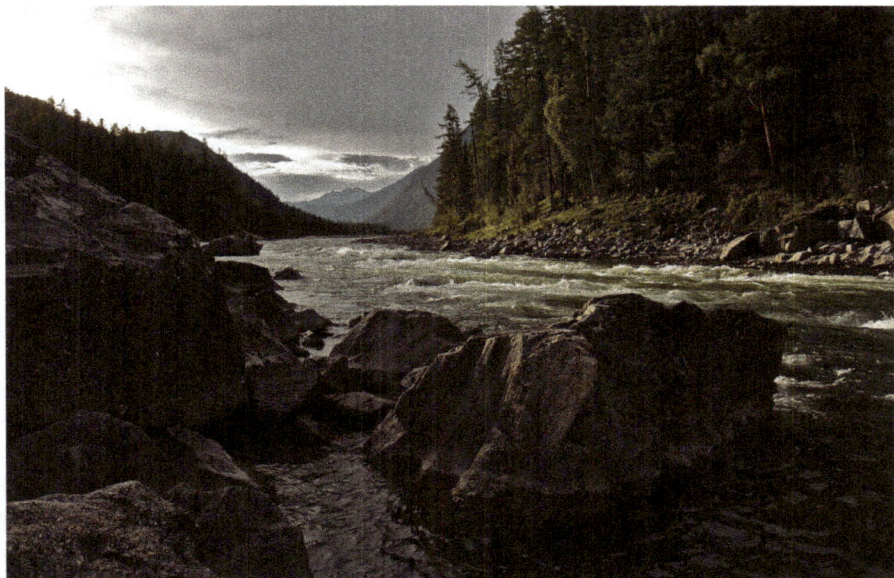

A brook puzzle for pupils to solve— When we have a load to carry we go slowly because we are obliged to; and the heavier the load, the slower we go. On the other hand, when we wish to run very swiftly we drop the load so as not to be weighted down; when college or high school boys run races in athletic games, they do not wear even their ordinary clothing, but dress as lightly as possible in trunks and tights; they also train severely so that they do not have to carry any more flesh on their bones than is necessary. How is it that in the case of a brook just opposite is true? The faster the brook runs, the more it can carry; and the heavier it becomes the faster it runs; and the faster it runs the more work it can do.

How a Brook Drops Its Load

THE brook is most discriminating in the way it takes up its burdens, and also in the way it lays them down. It, with quite superhuman wisdom, selects the lightest material first, leaving the heaviest to the last; and when depositing the load, it promptly drops the heaviest part first. And thus the flowing waters of the earth are eternally lifting, selecting, and sifting the soils on its surface.

The action of rain upon the surface of the ground is in itself an excellent lesson in erosion. If there is on a hillside a bit of bare ground which has been recently cultivated or graded, we can plainly see, after a heavy rain, where the finer material has been sorted out and carried away, leaving the larger gravel and stones. And if we examine the pools in the brook, we shall find deltas as well as many examples of the way the soil is sifted as it is dropped. The water of a rill flowing through pasture and meadow is clear, even after a hard rain. This is owing, not so much to the fact that the roots hold the banks of the brook firmly, as that the grass on the surface of the ground acts as a mulch and protects the soil from the erosive impact of the raindrops. On the other hand, and for a reverse reason, a rill through plowed ground is muddy. On a hillside, therefore, contour plowing is practiced—that is, plowing crosswise the hillside instead of up and down. When the furrow is carried crosswise, the water after showers can not dash away, carrying off in it all the finer and more fertile portions of the soil. There are many instances in our Southern States where this difference in the direction of plowing has saved or destroyed the fertility of hillside farms.

The little experiment suggested at the beginning of the following lesson, should show the pupils clearly the following points: It is through motion that water takes up soil and holds it in suspension. The tendency of still water is to drop all the load which it is carrying and it drops the heaviest part first. We find the pebbles at the bottom of the jar, the sand and gravel next, and the fine mud on top. The water may become perfectly clear in the jar and yet, when stirred a little, it will become roily again because of the movement. Every child who wades in a brook, knows that the edges and the still pools are more comfortable for the feet than is the center of the stream under the swift current. This is because, where the water is less swift at the sides, it deposits its mud and makes a soft bottom; while under the swifter part of the current, mud is washed away leaving the larger stones bare. For the same reason, the bottom of a stream crossing a level field is soft, because the silt, washed down from the hills by the swift current, is dropped when the waters come to a more quiet place. If, across a stony brook, the pupils can build a dam that will hold for two or three months in the fall or spring when the brook is flooded, they will be able to note that the stones will soon be more or less covered with soft mud; for the dam, stopping the current, causes the water to drop its load of silt. It would have to be a very recently made pool in a stream, which would not have a soft mud bottom. The water at times of flood is forced to the side of the streams in eddies, and its current is thus checked, and its load of mud dropped.

It should be noted that at points where the brook is narrowest the current is swiftest, and where the current is swiftest the bottom is more stony. Also, where there is a bend in the stream the brook digs

Where a book is thinnest it runs the fastest

deeper into the bank where it strikes the curve, and much of the soil thus washed out is removed to the other side of the stream where the current is very slow, and there is dropped. (See *Introduction to Physical Geography*, Gilbert and Brigham, pp. 51 and 52.) If possible, note that where a muddy stream empties into a pond or lake, the waters of the latter are made roily for some distance out, but beyond this the water remains clear. The pupils should be made to see that the swift current of the brook is checked when its waters empty into a pond or lake, and because of this they drop their load. This happens year after year, and a point extending out into the lake or pond is thus built up. In this manner the great river deltas are formed.

References— *The Brook Book*, Mary Rogers Miller; *Brooks and Brook Basins*, Frye; *Up and Down the Brooks*, Bamford; *Physical Geography*, Tarr; *Introduction to Physical Geography*, Gilbert and Brigham.

LESSON

Leading thought— The brook carries its load only when it is flowing rapidly. As soon as the current is checked, it drops the larger stones and gravel first and then the finer sediment. It is thus that deltas are built up where streams empty into lakes and ponds.

Method— Study the rills made in freshly graded soil directly after a heavy rain. Ask the pupils individually to make observations on the flooded brook.

Experiment— Take a glass fruit jar nearly full of water from the brook, add gravel and small stones from the bed of the brook, sand from its borders and mud from its quiet pools. Have it brought into the schoolroom, and shake it thoroughly. Then place in a window and ask the pupils to observe the following things:

(a) Does the mud begin to settle while the water is in motion; that is, while it is being shaken?

(b) As soon as it is quiet, does the settling process begin?

(c) Which settles first—the pebbles, the sand or the mud? Which settles on top—that is, which settles last?

(d) Notice that as long as the water is in the least roily, it means that the soil in it has not all settled; if the water is disturbed even a little it becomes roily again, which means that as soon as the water is in motion it takes up its load.

Observations—

1. Where is the current swiftest, in the middle or at the side of the stream?

2. What is the difference, in the bottom of the brook, between the place below the swift current and the edges? That is, if you were wading in the brook, where would it be more comfortable for your feet—at the sides or in the swiftest part of the current? Why?

3. Does the brook have a more stony bed where it flows down a hillside than where flowing through a level place?

4. Place a dam across your brook where the bottom is stony, and note how soon it will have a soft mud bottom. Why is this?

5. Can you find a still pool in your brook that has not a soft, muddy bottom? Why is this?

6. Does the brook flow more swiftly in the steep and narrow places than in the wide portions and where it is dammed?

7. Do you think if water, flowing swiftly and carrying a load of mud, were to come to a wider or more level place, like a pool or mill-

pond dam, that it would drop some of its load? Why?

8. If the water flows less swiftly along the edges than in the middle, would this make the bottom below softer and more comfortable to the feet than where the current is swiftest? If so, why?

9. If you can see the place where a brook empties into a pond or lake, how does it make the waters of the latter look after a storm? What is the water of the brook doing to give this appearance, and why?

10. What becomes of the soil dropped by the brook as it enters a pond or lake? Do you know of any points of land extending out into a lake or pond where the stream enters it? What is a stream delta?

"In the bottom of the valley is a brook that saunters between oozing banks. It falls over stones and dips under fences. It marks an open place on the face of the earth, and the trees and soft herbs bend their branches into the sunlight. The hangbird swings her nest over it. Mossy logs are crumbling into it. There are still pools where the minnows play. The brook runs away and away into the forest. As a boy I explored it but never found its source. It came somewhere from the Beyond and its name was Mystery.

The mystery of this brook was its changing moods. It had its own way of recording the passing of the weeks and months. I remember never to have seen it twice in the same mood, nor to have got the same lesson from it on two successive days: yet, with all its variety, it always left that same feeling of mystery and that same vague longing to follow to its source and to know the great world that I was sure must lie beyond. I felt that the brook was greater and wiser than I. It became my teacher. I wondered how it knew when March came, and why its round of life recurred so regularly with the returning seasons. I remember that I was anxious for the spring to come, that I might see it again. I longed for the earthy smell when the snow settled away and left bare brown margins along its banks. I watched for the suckers that came up from the river to spawn. I made a note when the first frog peeped. I waited for the unfolding spray to soften the bare trunks. I watched the greening of the banks and looked eagerly for the bluebird when I heard his curling note somewhere high in the air."

—"THE NATURE-STUDY IDEA," L. H. BAILEY.

A snowflake

Crystal Growth

TEACHER'S STORY

TO watch the growth of a crystal is to witness a miracle; involuntarily we stand in awe before it, as a proof that of all truths mathematics is the most divine and inherent in the universe. The teacher will fail to make the best use of this lesson if she does not reveal to the child through it something of the marvel of crystal growth.

That a substance which has been dissolved in water should, when the water evaporates, assemble its particles in solid form of a certain shape, with its plane surfaces set exactly at certain angles one to another, always the same whether the crystal be large or small, is quite beyond our understanding. Perhaps it is no more miraculous than the growth of living beings, but it seems so. The fact that when an imperfect crystal, unfinished or broken, is placed in water which is saturated with the same substance, it will be built out and made perfect, shows a law of growth so exquisitely exemplified as to again make us glad to be a part of a universe so perfectly governed. Moreover, when crystals show a variation in numbers of angles and planes it is merely a matter of division or multiplication. A snow crystal is a six-rayed star, yet sometimes it has three rays.

The window-sill of a schoolroom may be a place for the working of greater wonders than those claimed by the astrologists of old, when they transmuted baser metals to gold and worthless stones to diamonds. It may be a place where strings of gems are made before the wondering eyes of the children; gems fit to make necklaces for any naiad of the brook or oread of the caves.

It adds much to the interest of this lesson if different colored substances are used for the forming of the crystals. Blue vitriol, potassium bichromate, and alum give beautiful crystals, contrasting in shape as well as in colors.

Copper sulphate and blue vitriol are two names for one substance; it is a poison when taken internally and, therefore, it is best for the teacher to carry on the experiment before the pupils instead of trusting the substance to them indiscriminately. Blue vitriol forms an exquisitely beautiful blue crystal, which is lozenge-shaped with oblique edges. Often, as purchased from the drug store, we find it in the form of rather large, broken, or imperfect crystals. One of the pretty experiments is to place some of these broken crystals in a saucer containing a saturated solution of the vitriol, and note that they straightway assert crystal nature by building out the broken places, and growing into perfect crystals. Blue vitriol is used much in the dying and in the printing of cotton and linen cloths. It has quite wonderful preservative qualities; if either animal or vegetable tissues are permeated by it they will remain dry and unchanged.

Potassium bichromate is also a poison and, therefore, the teacher should make the solution in the presence of the class. It forms orange-red crystals, more or less needle-shaped. It crystallizes so readily that if one drop of the solution be placed on a saucer the pupils may see the formation of the crystals by watching it for a few moments through a lens.

The common alum we buy in crystal form, however, is very much broken. Its crystals are eight-sided and pretty. Alum is widely used in dyes, in medicines, and in many other ways. It is very astringent, as every child knows who has tried to eat it, and has found the lips and tongue much puckered thereby.

Although we are more familiar with crystals formed from substances dissolved in water, yet there are some minerals, like iron, which crystallize only when they are melted by heat; and there are other crystals, like the snow, which are formed from vapor. Thus, substances must be molten hot, or dissolved in a liquid, or in form of gas, in order to grow into crystals.

LESSON

Leading thought— Different substances when dissolved in water will re-form as crystals; each substance forms crystals of its own peculiar color and shape.

Method— Take three test tubes, long vials or clear bottles. Fill one with a solution made by dissolving one part of blue vitriol in three parts of water; fill another by dissolving one part of bichromate of potash with twenty-five parts of water; fill another with one part of alum in three parts of water. Suspend from the mouth of each test tube or vial, a piece of white twine, the upper end tied to a tooth pick, which is placed across the mouth of the vial; the other end should reach the bottom of the vial. If necessary, tie a pebble to the lower end so that it

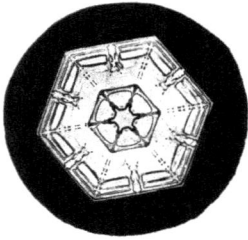

A snow crystal

will hang straight. Place the bottles on the window sill of the schoolroom, where the children may observe what is happening. Allow them to stand for a time, until the string in each case is encrusted with crystals; then pull out the string and the crystals. Dry them with a blotter, and let the children observe them closely. Care should be taken to prevent the children from trying to eat these beautiful crystals, by telling them that the red and blue crystals are poisonous.

Observations—

1. In which bottle did the crystals form first? Which string is the heaviest with the crystals?

2. What was the color of the water in which the blue vitriol was dissolved? Is it as brilliant in color now as it was when it was first made? Do you think that the growth of the crystals took away from the blue material of the water? Look at the blue vitriol crystals with a lens, and describe their shape. Are the shapes of the large crystals of the vitriol the same as those of the small ones?

3. What is the shape of the crystals of the potassium bichromate? What is the color? Are these crystals as large as those of the blue vitriol or of the alum?

4. What shapes do you find among the crystals of alum?

5. Do you think that vitriol and potassium bichromate and alum will, under favorable circumstances, always form each its own shape of crystal wherever it occurs in the world? Do you think crystals could be formed without the aid of water?

6. How many kinds of crystals do you know? What is rock candy? Do you think you could make a string of rock candy if you dissolved sugar in water and placed a string in it?

A salt crystal at the Devil's Golf Course

Salt

TEACHER'S STORY

A "SATURATED SOLUTION" is an uninspiring term to one not chemically trained; and yet it merely means water which holds as much as it can take of the dissolved substance; if the water is hot, it dissolves more of most substances. To make a saturated solution of salt we need two parts of salt or a little more, for good measure, to five parts of water; the water should be stirred until it will take up no more salt.

A slip of paper placed in a saucer of this solution will prove a resting place for the crystals as they form. In about two days the miracle will be working, and the pupils should now and then observe its progress. Those saucers set in a draft or in a warm place will show crystals sooner than others, but the crystals will be smaller; for the faster a crystal grows, the smaller is its stature. If the water evaporates rapidly, the crystals are smaller, because so many crystals are started which do not have material for large growth. When the water is evaporated, to appreciate the beauty of the crystals we should look at them with a lens or microscope. Each crystal is a beautiful little cube, often with a pyramid-shaped depression in each face or side. After the pupils have seen these crystals, the story of where salt is found should be told them.

Salt is obtained by two methods: by mining large deposits of rock salt, and by evaporating water containing a strong solution of salt. The oldest salt works in this country are in Syracuse, New York, where the salt comes from salt springs which were famous among the American Indians. At Ithaca, N. Y., the salt deposits are about 2000 feet below the surface of the earth. Water is forced down into the stratum of rock, which was evidently once the bottom of a briny sea; the water dissolves the salt, and it is then pumped up to the surface and evaporated, leaving the salt in crystals. In Michigan and Louisiana there are other large salt works of a similar character. The largest salt mines in the world are those in Poland, which have been used for hundreds of years. In these mines there are fifty miles of corridors, and the salt has been carved into beautiful chambers with statues and other decorations, all cut from the solid salt. One of these chambers represents a chapel beautifully ornamented.

When the United States was first settled, salt was brought over from England; but this was so expensive that people could not afford it and they soon began to make their own salt by evaporating sea water in kettles on the beach. In those countries where it is scarce, salt is said to be literally worth its weight in gold. The necessity for salt to preserve the health of both people and animals has tempted the governments of some countries to place a special tax upon it; in Italy, especially, the poor people suffer greatly on account of the high price of salt from this cause.

Salt lakes are found in natural basins of arid lands, and are always without outlets. The water which runs in escapes by evaporation, but the salt it brings cannot escape, and accumulates. A salt lick is a place where salt is found on the surface of the earth, usually near a salt spring. Animals will travel a long distance to visit a salt lick which gained its name through their attentions.

LESSON

Leading thought— Salt dissolves in water, and as the water evaporates the salt appears in beautiful crystals.

Method— Let each pupil, if possible, have a cup and saucer, a square of paper small enough to go into the saucer, some salt and wa-

Salt crystals

ter. Let each pupil take five teaspoonfuls of water and add to this two spoonfuls of salt, stirring the mixture until dissolved. When the water will take no more salt let each pupil write his name and the date on the square of paper, lay it in the saucer, pressing it down beneath the surface. Let some place their saucers in a warm place, others where they may be kept cool, and others in a draft. If it is impossible for each pupil to have a saucer, two or three pupils may be selected to perform the experiments.

Observations—

1. When you pour the salt into the water, what becomes of it? How do you know when the water will hold no more salt?

2. After a saucer, filled with the salt water, stands exposed to the air for several days, what becomes of the water? From which saucers did the water evaporate fastest—those in the warm places, or those in the cold? In which did the crystals form first?

3. Which saucers contained the largest crystals—those from which the water evaporated first, or those from which it evaporated more slowly?

4. Could you see how the crystals began? What is the shape of the perfect salt crystal? Do the smallest crystals have the same shape as the largest ones?

5. What happens to people who cannot get salt to eat?

6. How is dairy salt and table salt obtained? What is rock salt? What are salt licks? Where are the salt mines found? Why is the ocean called "the briny deep?"

7. Name and locate the salt lakes. Why are lakes salt?

How To Study Minerals

LESSON

MANY children are naturally interested in stones. I once knew two children, aged seven and five, who could invariably select the boulders and pebbles of metamorphic rock in the region about Ithaca. They also could tell, when the pebbles were broken, which parts were quartz and which mica. They had incidentally asked about one of these stones, and I had told them the story of the glacial period and how these stones were torn away from the mountains in Canada and brought down by ice and dropped in Ithaca. It was a story they liked, and their interest in these granite voyagers was always a delightful element of our walks in the field.

For the pupils in the elementary grades it seems best to limit the study of minerals to those which make up our granite and common rocks. In order to teach about these minerals well, the teacher should have at least one set of labelled specimens. Such a collection may be obtained from Edward E. Howell, 612 17th St., N. W., Washington, D. C., and also from Ward's Natural Science Establishment, College

Avenue, Rochester, N. Y. These collections vary in number of specimens and price from one to two dollars and are excellent. The teacher should have one or two perfect crystals of quartz, feldspar and calcite. An excellent practice for a boy is to copy these crystals in wood for the use of the teacher.

The physical characteristics used in identifying minerals are briefly as follows:

1. *Form.* This may be crystalline, which shows the shape of the crystals definitely; granular, like marble, the grains having the internal structure, but not the external form, of crystals; compact, which is without crystalline form, as limestone or flint.

2. *Color.*

3. *Luster or shine*, which may be glassy like quartz; pearly like the inside of a shell; silky like asbestos; dull; or metallic like gold.

4. *Hardness* or resistance to scratching, thus: Easily scratched with the finger nail; cannot be scratched by the finger nail; easily scratched with steel; with difficulty scratched with steel; not to be scratched by steel. A pocket knife is usually the implement used for scratching.

A rock collection

23

Quartz

THERE is in the Cornell University Museum a great quartz crystal, a six-sided prism several inches in thickness. One-half of it is muddy and the other half clear, transparent and beautiful. The professor in charge, who has the imagination necessary to the expert crystallographer, said to his class: "This crystal was begun under conditions which made it cloudy; then something happened, perhaps some cataclysm that changed all the conditions around the half-grown crystal, and it may have lain a hundred or a thousand years unfinished, when, some other change occurring, there came about conditions which permitted it to resume growth, and the work began again exactly where it was left off, the shaft being perfected even to its six-sided pyramidal tip." And ever afterwards that crystal, half clouded and half clear, remained in the minds of his pupils as a witness of the eternal endurance of the laws which govern the growth of crystals.

Quartz is the least destructible and is one of the most abundant materials in the crust of the earth as we know it. It is made up of two

24

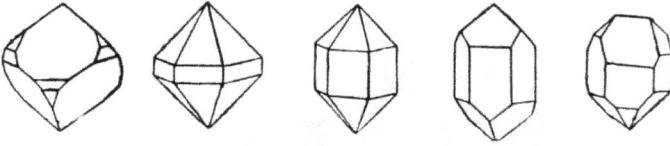

Forms of quartz crystals

elements chemically united—the solid silicon and the gas oxygen. It is the chief material of sand and sandstones, and it occurs, mixed with grains of other minerals, in granite, gneiss, and many lavas; it also occurs in thick masses or sheets, and sometimes in crystals ornamenting the walls of cavities in the rocks. Subterranean waters often contain a small amount of silica, the substance of quartz, in solution; from such solutions it may be deposited in fissures or cracks in the rock, thus forming bodies called "veins." Other materials are often deposited at the same time, and in this way the ores of the precious metals came to be associated with quartz. Sometimes quartz is deposited from hot springs or geysers, forming a spongy substance called sinter. In this case, some of the water is combined with the quartz, making what is called opal. Quartz crystal will cut glass.

Quartz occurs in many varieties: (a) In crystals like glass. If colorless and transparent it is called rock crystal; if smoky brown, it is called smoky quartz; if purple, amethyst. (b) In crystals, glassy but not transparent. If white, it is milky quartz; if pink, rose quartz. (c) As a compact crystalline structure without luster, waxy or dull, opaque or translucent, when polished. If bright red, it is carnelian; if brownish red, sard; if in various colors in bands, agate; if in horizontal layers, onyx; if dull red or brown, jasper; if green with red spots, bloodstone; if smoky or gray, breaking with small, shell-like or conchoidal fractures, flint.

Rock crystals are used in jewelry and especially are made to imitate diamonds. The amethyst is much prized as a semi-precious stone. Carnelian, bloodstone and agate are also used in jewelry; agate is used also in making many ornamental objects, and to make little mortars and pestles for grinding hard substances.

One of the marvels of the world is the petrified forest of Arizona, now set aside by the government as a national reserve. Great trees have been changed to agate and flint, the silica being substituted for

Purple quartz is called amethyst

the tissues of the wood so that the texture is preserved though the material is changed.

When our country was first settled, flint was used to start fires by striking it with steel and letting the sparks fly into dry, fine material, called tinder. It was also used in guns before the invention of cartridges, and the guns were called flintlocks. The Indians used flint to make hatchets and for tips to their arrows. The making of flint implements dates far back into prehistoric times; it was probably one of the first steps upward which man achieved in his long, hard climb from a level with the brute creation to the heights attained by our present civilization.

Quartz sand is used in making glass. It is melted with soda or potash or lead, and the glass varies in hardness according to the minerals added. Quartz is also used for sandpaper and glass paper; and ground to a fine powder, it is combined with Japans and oils and used as a finish for wood surfaces. Mineral wool is made from the slag refuse of furnaces where glass is made, and is used for rat-proof and fireproof padding for the walls of houses. Quartz combined with sodium or potassium and water, forms a liquid called water-glass, which is used for waterproof surfaces; it is also fireproof to a certain degree.

Smoky quartz

Water-glass is the best substance in which to preserve eggs; one part of commercial water-glass to ten parts of water makes a proper solution for this purpose.

LESSON

Leading thought— Quartz is one of the most common of minerals. It occurs in many forms. As a crystal it is six-sided, and the ends terminate in a six-sided pyramid. It is very hard and will scratch and cut glass. When broken, it has a glassy luster and it does not break smoothly but shows an uneven surface.

Method— The pupils should have before them as many varieties of quartz as possible; at least they should have rock crystal, amethyst, rose and smoky quartz and flint.

Observations—

1. What is the shape of quartz crystals? Are the sides all of the same size? Has the pyramid-shaped end the same number of plane surfaces as the sides?

2. What is the luster of quartz? Is this luster the same in all the different colored kinds of quartz?

3. Can you scratch quartz with the point of a knife? Can you scratch glass with a corner or piece of the quartz? Can you cut glass with quartz?

4. Describe the following kinds of quartz and their uses: amethyst, agate, flint.

5. How many varieties of quartz do you know? What has quartz to do with the petrified forests of Arizona?

A superb feldspar crystal

Feldspar

Teacher's Story

WE most commonly see feldspar as the pinkish portion of granite. This does not mean that feldspar is always pink, for it may be the lime-soda form known as labradorite, which is dark gray, brown or greenish brown, or white; or it may be the soda-lime feldspar called oligoclase, which is grayish green, grayish white, or white; but the most common feldspar of all is the potash feldspar—orthoclase—which may be white, nearly transparent, or pinkish. Orthoclase is different from other feldspars in that, when it splits, its plane surfaces form right angles. Feldspar is next in the scale of hardness to quartz, and will with effort and perseverance scratch glass but will not cut it; it can be scratched with a steel point. Its luster is glassy and often somewhat pearly.

Maine leads all other states in the production of feldspar. It is quarried and crushed and ground to powder, as fine as flour, to make the clay from which china and all kinds of pottery are made. Our clayey soils are made chiefly from the potash feldspar which is weathered to fine dust. Kaolin, which has been used so extensively in making the finest porcelain, is the purest of all clays, and is formed of weathered feldspar; floor tiling and sewer pipes are also made from ground feldspar. Moonstone is clean, soda-lime feldspar, whitish in color and with a reflection something like an opal.

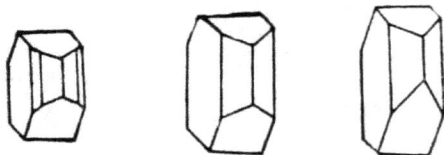

Forms of feldspar crystals

LESSON

Leading thought— Feldspar is about five times as common as quartz. The crystal is obliquely brick-shaped, and when broken splits in two directions at right angles to each other. It is next in hardness to quartz, and will scratch glass but will not cut it.

Method— If possible, have the common feldspar (orthoclase), the soda-lime feldspar (oligoclase) and the lime-soda feldspar (labradorite).

Observations—

1. What is the shape of the feldspar crystal?

2. What colors are your specimens of feldspar? How many kinds have you?

3. What is the luster of feldspar?

4. Can you scratch feldspar with the point of a knife? Can you scratch it with quartz? Can you scratch glass with it?

5. When you scratch feldspar with steel what is the color of the streak left upon it?

6. If feldspar is broken, does it break along certain lines, leaving smooth faces? At what angles do these smooth faces stand to each other?

7. How can you tell feldspar from quartz? Write a comparison of feldspar and quartz, giving clearly the characteristics of both.

8. Hunt over the pebbles found in a sand-bank. Which ones are quartz? Do you find any of feldspar?

9. When there is so much more feldspar than quartz in the earth's crust, why is there so much more quartz than feldspar in sand?

Mica has visible sheets

Mica

TEACHER'S STORY

THE mica crystal when perfect is a flat crystal with six straight edges. These crystals separate in thin layers parallel with the base. In color mica varies, through shades of brown, from a pale smoked pearl to black. Its luster is pearly, and it can be scratched with the thumb nail. Its distinguishing characteristic is that the thin layers into which it splits bend without breaking and endure great heat.

Mica was used in antiquity for windows. Because it is transparent and not affected by heat, it is used in the doors of stoves and furnaces and for lamp chimneys. Its strength makes it of use for automobile goggles. Diamond dust is powdered mica, as is also the artificial snow scattered over cotton batting for the decoration of Christmas trees. When ground finely, it is used as an absorbent for nitroglycerine in the manufacture of dynamite.

Mica mines are scarce in this country. There is an interesting one in North Carolina which had evidently been worked centuries before the advent of the white man in America. There are other mica mines in New Hampshire and Canada. The entire production of this mineral in the United States for the year 1908, was valued at a little more than a quarter of a million dollars. Nearly all

Dark Mica

of this output was used in the electrical industries, since mica is one of the best insulating materials known.

LESSON

Leading thought— Mica is a crystal which flakes off in thin scales parallel with the base of the crystal. We rarely see a complete mica crystal but simply the thin plates which have split off. The ordinary mica is light colored, but there is a black form.

Method— If it is not possible to obtain a mica crystal, get a thick piece of mica which the pupils may split off into layers.

Observations—

1. Describe your piece of mica. Pull off a layer with the point of your knife. See if you can separate this layer into two layers or more.

2. Can you see through mica? Can you bend it? Does it break easily? What is the color of your specimen? What is its luster? Can you cut it with a knife? Can you scratch it with the thumb nail? What color is the streak left by scratching it with steel?

3. What are some of the uses of mica? How is it especially fitted for some uses?

4. Write a theme on how and where mica is obtained.

Granite peaks in Huangshan, China

Granite

TEACHER'S STORY

IN granite, the quartz may be detected by its fracture which is always conchoidal and never flat; that is, it has no cleavage planes. It is usually white or smoky, and is glassy in luster. It cannot be scratched with a knife. The feldspar is usually whitish or flesh-colored and the smooth surface of its cleavage planes shines brilliantly as the light strikes upon it; it can be scratched with a knife but this requires effort. The mica is in pearly scales, sometimes whitish and sometimes black. The scales of these mica particles may be lifted off with a knife, and it may thus be distinguished. If there are black particles in the granite which do not separate, like the mica, into thin layers, they probably consist of hornblende.

Granite is used extensively for building purposes and for monuments. It is a very durable stone but in the northeastern United States where there is much rain and cold weather, the stone decays. Mica is

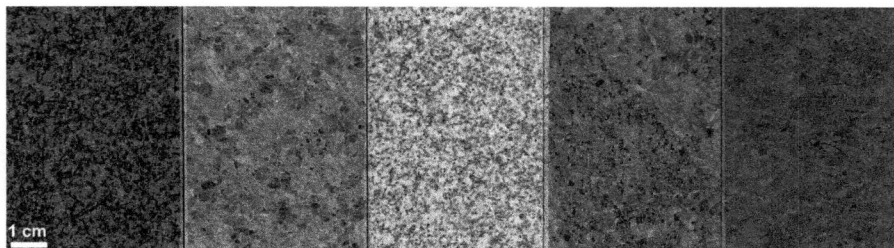
Various shades of granite

the weakest, hornblende next, and feldspar is next to quartz, the strongest constituent of granite. Water permeates the mica, hornblende, feldspar and sometimes the quartz, and by its expansion in freezing causes the stone to crumble. The reason why polished granite endures better than the rough finished, is that the smooth surface gives less opportunity for the water to lodge and freeze. When the weathered granite is cut up into small particles by the waters of streams, they are sifted and all the parts which are soluble are carried off, leaving a sand composed of quartz and mica, which are insoluble. This sand is washed by streams into lakes, and then is dropped to the bottom; if enough is thus carried and dropped, it forms sandstone rock. All of our sandstones used for building purposes were thus laid down.

Cleopatra's Needle, which stood for thousands of years in the dry climate of Egypt, soon commenced to weather and crumble when placed in Central Park, N. Y. This shaft has a most interesting history. It was quarried near Assuan, in the most famous of all granite quarries of ancient Egypt. It was cut as a solid shaft in the quarry and carried down the Nile River for 500 miles—an engineering feat which would be hard to accomplish to-day, with all our modern appliances. It was one of the obelisks that graced the ancient city of On, later called Heliopolis, situated on a plateau near the present city of Cairo; On was the city where Moses was born and reared. There is still standing where it was first placed as a part of a magnificent temple, the temple a part of a magnificent city, one of these obelisks. It now stands alone in the middle of a great fertile plain, which is vividly green with growing crops; a road shaded by tamarisk and lebbakh trees leads to it; nearby is a sakiah, creaking as the blindfolded bullock walks around and around, turning the wheel that lifts the chain of buckets from the well to irrigate the crops; and a hooded crow, whose ancestors were contemporaries of its erection, caws hoarsely as it alights on the beautiful apex of this ancient shaft, which has stood there nearly four thousand years and has seen a great city go down to dust to fertilize a grassy plain.

LESSON

Leading thought— Granite is composed of feldspar, quartz and mica, and often contains hornblende.

Method— Specimens of coarse granite and a pocket knife are needed.

Observations—

Cleopatra's Needle, a granite obelisk

1. What minerals do you find in granite? How can you tell what these minerals are? Look at the granite with a lens. How can you tell the quartz from feldspar? Take a knife and scratch the two. Can you tell them apart in that way? How can you tell the mica? How can you tell the hornblende?

2. What buildings have you seen made of granite? What monuments have you seen made from it?

3. Which mineral in granite is especially affected by water? Which remains unharmed the longest?

4. What is weathering? Mention some of the characteristics of weathering. Why does the rough-finished granite weather sooner than that which is polished?

5. Examine some sand with a lens. What mineral do you find present in it in the greatest quantity?

6. Write the story of the Cleopatra's Needle in Central Park, New York City.

Calcite, Marble, and Limestone

Calcite

CALC spar, or calcium carbonate, is a mineral and is the material of which marble, limestone and chalk are made. The faces of the calcite crystal are always arranged in groups of three or multiples of three—a three-sided pyramid or two pyramids joined base to base. The pyramids may be obtuse or acute. When acute and formed of three pairs of faces, the crystals are called dog-tooth spar. The crystals appear in a great variety of forms, but they all have the common quality of splitting readily in three directions, the fragments resembling cubes which are oblique instead of rectangular. When these cleaved, or split pieces, are transparent, they are called Iceland spar. When an object is viewed through Iceland spar it least one-quarter inch thick, it appears double. The calcite crystal is often transparent with a slight yellowish tinge, but it also shows other colors; and it has a slightly cloudy or slightly pearly or almost glassy luster like feldspar. It is easily scratched with a knife and will not scratch glass. If a drop of strong vinegar or weak hydrochloric acid falls upon it, it will effervesce.

Limestone—so called because it is burned to make quicklime—was formed on the bottoms of oceans; its substance came chiefly from the skeletons of corals and the shells of other sea creatures, since sea-shells and coral stems are pure calcium carbonate in composition. In the water, the shells and corals were broken down, and then deposited in layers on the bottom of the sea. So wherever we find limestone, we know that there was once the bottom of a great sea. Such layers of limestone are now being deposited off the shores of Florida, where corals grow in great abundance. Limestone is used extensively for building purposes, and in most climates is very durable. The great

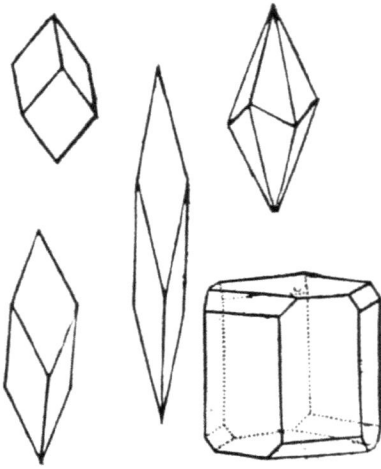
Forms of calcite crystals

pyramids of Egypt are of limestone. It is not a good material for making roads, since it is so soft that it wears out readily, making a fine easily-blown dust. It is slowly dissolved in water, especially if the water be acid; thus, in limestone regions, there are caves where the water has dissolved out the rock; and attached to their roofs and piled upon their floors may be large icicle-shaped stalactites and stalagmites, which were made by the lime-bearing water dripping down and evaporating, leaving its burden in crystals behind it. When the roof of a cave falls in, the cavity thus made is called a sink hole and is often dangerous. The famous Natural Bridge in Virginia is all that is left of what was once the roof of such a cavern. The water in limestone regions is always hard, because of the lime which it holds in solution; and in such regions the streams usually have no silt, but have clean bottoms; moreover, the springs are likely to become contaminated because the water has run through long caves instead of filtering through sand.

Chalk is similar in origin to limestone; it is made up of the shells of minute sea creatures, so small that we can only see them with the aid of a microscope. Try and think how many years it must have required for the shells of such tiny beings to build up the beds which make the great chalk cliffs of England!

Marble is formed inside of the earth from limestone, under the influence of heat and pressure; it differs from limestone chiefly in that the grains are of crystalline structure, and are larger; it is usually white or gray in color, and sometimes is found in differing colors. At Cadiz in California, marble is found showing twenty or more quite different colors. The most famous marbles are the Carrara of Italy, the Parian from the Island of Paros, and the Pentelican from the mountain of that name near Athens. The reason why these marbles are so famous is that in ancient times sculptors carved beautiful statues from them, and the architects used them for building magnificent temples. The principal marble deposits in the United States are in Vermont, Georgia, Tennessee and California. Marble deteriorates when it is exposed to air which is filled with smoke and gases. It is also used

ZYANCE (CC BY-SA 2.5)

Raw marble

to make lime. When either marble or limestone is heated very hot, it separates into two parts, one of which is lime, and the other carbonic acid gas—the same that is used for charging soda-water fountains.

LESSON

Leading thought— Calcite or calc spar is formed more than half of lime. The best known forms of its crystals are cubelike, but instead of having twelve right-angled edges, the sides are lozenge-shaped, and are set together with six obtuse angles and six acute. Dog-tooth spar is one form of calcite crystal. Limestone is a solid form of calcite. Marble is granular limestone which shows the broken crystals of calcite. Chalk is very fine, pulverized calcite.

Method— Specimens of dog-tooth spar, limestone, marble, shells of oysters or other sea creatures and coral should be provided for this lesson; also a bottle of dilute hydrochloric acid, and a piece of glass tubing about six inches long with which to drop the acid on the stones. Some strong vinegar will do instead of the acid.

Observations—

1. What is the form of the calcite crystal? What is the luster of the crystal? Is it the same as the inside of sea-shells? Will calcite scratch glass? Can you scratch it with a knife? What happens to calcite if you put a drop of weak hydrochloric acid upon it?

2. Is marble made up of crystals? Examine it with a lens to see. What is its color? Have you seen marble of other colors than white? Do you know the reason why marble is sometimes clouded and streaked?

3. Put a drop of weak hydrochloric acid on the marble. What happens?

4. What are the uses of marble? What have you ever seen made from marble? Why is it used for sculpture? What famous statues have you seen which were made of marble? Name some of the famous ancient marble buildings.

Limestone is used as a building material

5. Test a piece of limestone for hardness. Can you scratch it with a knife? Is it as soft as marble? Put on it a drop of acid. Does it effervesce? If there are any fossils in your piece of limestone, test them with acid and see if they will effervesce. Any other mineral that you have which will effervesce when touched with acid, is probably some form of calcite.

6. Are there any buildings in your town made of limestone? How do you know the stone is limestone? Where was it obtained? Is it affected by the weather?

7. Is limestone a good material for making or mending roads? Give a reason.

8. Why is water in limestone regions hard? Why are limestone regions likely to have caves within the rocks? How are stalactites and stalagmites formed in caves? What are sink holes? How are they formed? In what county of your state is limestone found?

9. How is the lime which is used for plastering houses made?

10. Write a theme on how the chalk rocks are made.

The white cliffs of dover are made of chalk, a very fine limestone

11. Test a shell with acid; test a piece of coral with acid. How does it happen that these, which were once a part of living creatures, are now limestone? Of what are our own bones made?

"A great chapter in the history of the world is written in the chalk. Few passages in the history of man can be supported by such an overwhelming mass of direct and indirect evidence as that which testifies to the truth of the fragment of the history of the globe, which I hope to enable you to read, with your own eyes, to-night. Let me add, that few chapters of human history have a more profound significance for ourselves. I weigh my words well when I assert, that the man who should know the true history of the bit of chalk which every carpenter carries about in his breeches-pocket, though ignorant of all other history, is likely, if he will think his knowledge out to its ultimate results, to have a truer, and therefore a better, conception of this wonderful universe, and of man's relation to it, than the most learned student who is deep-read in the records of humanity and ignorant of those of Nature."

"During the chalk period, or 'cretaceous epoch,' not one of the present great physical features of the globe was in existence. Our great mountain ranges, Pyrenees, Alps, Himalayas, Andes, have all been upheaved since the chalk was deposited, and the cretaceous sea flowed over the sites of Sinai and Ararat. All this is certain, because rocks of cretaceous or still later date, have shared in the elevatory movements which give rise to these mountain chains; and may be found perched up, in some cases, many thousand feet high upon their flanks."

—Thomas Huxley.

The Magnet

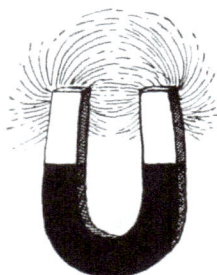

NTIL comparatively recent times, the power of the magnet was so inexplicable that it was regarded as the working of magic. The tale of the Great Black Mountain Island magnet described in the "Arabian Nights Entertainments"—the story of the island that pulled the nails from passing ships and thus wrecked them—was believed by the mariners of the Middle Ages. Professor George L. Burr assures me that this mountain of lodestone and the fear which it inspired were potent factors in the development of Medieval navigation. Even yet, with all our scientific knowledge, the magnet is a mystery. We know what it does, but we do not know what it is. That a force unseen by us is flowing off the ends of a bar magnet, the force flowing from one end attracted to the force flowing from the other and repellent to a force similar to itself, we perceive clearly. We also know that there is less of this force at a point in the magnet half-way between the poles; and we know that the force of the magnet acts more strongly if we offer it more surface to act upon, as is shown in the experiment in drawing a needle to a magnet by trying to attract it first at its point and then along its length. That this force extends out beyond the ends of the magnet, the child likes to demonstrate by seeing across how wide a space the magnet, without touching the objects, can draw to it iron filings or tacks. That the magnet can impart this force to iron objects is demonstrated with curious interest, as the child takes up a chain of tacks at the end of the magnet; and yet the tacks when removed from the magnet have no such power of cohesion. That some magnets are stronger than others is shown in the favorite game of "stealing tacks," the stronger magnet taking them away from the weaker; it can also be demonstrated by a competition between magnets, noting how many tacks each will hold.

One of the most interesting things about a magnet is that like poles repel and opposite poles attract each other. How hard must we pull to separate two magnets that have the south pole of one against the north pole of the other! Even more interesting is the repellent power of two similar poles, which is shown by approaching a suspended magnetized needle with a magnet. These attractive and repellent forces are most interestingly demonstrated by the experiment in question 13 of the lesson. These needles floating on cork join the magnet or flee from it, according to which pole is presented to them.

Not only does this power reside in the magnet, but it can be imparted to other objects of iron and steel. By rubbing one pole of the magnet over a needle several times, always in the same direction, the needle becomes a magnet. If we suspend such a needle by a bit of thread from its center, and the needle is not affected by the nearness of a magnet, it will soon arrange itself nearly north and south. It is well to thrust the needle through a cork, so it will hang horizontally, and then suspend the cork by a thread. The magnetized needle will not point exactly north, for the magnet poles of the earth do not quite coincide with the poles of the earth's axis.

The direction assumed by the magnetized needle may be explained

by the fact that the earth is a great magnet, but the south pole of the great earth magnet lies near the North Pole of the earth. Thus, a magnet on the earth's surface, if allowed to move freely, will turn its north pole toward the south pole of the great earth magnet. Then, we might ask, why not call the earth's magnetic pole that lies nearest our North Pole its north magnetic pole? That is merely a matter of convenience for us. We see that the compass needle points north and south, and the arm of the needle which points north we conveniently call its north pole.

The above experiment with a suspended needle shows how the mariner's compass is made. This most useful instrument is said to have been invented by the Chinese, at least 1400 B. C., and perhaps even longer ago. It was used by them to guide armies over the great plains, and the needle was made of lodestone. The compass was introduced into Europe about 1300 A. D., and has been used by mariners ever since. To "box the compass" is to tell all the points on the compass dial, and is an exercise which the children will enjoy.

We are able to tell the direction of the lines of force flowing from a magnet, by placing fine iron filings on a pane of glass or a sheet of paper and holding close beneath one or both poles of a magnet; instantly the filings assume certain lines. If the two ends of a horseshoe magnet are used, we can see the direction of the lines of force that flow from one pole to the other., It is supposed that these lines of magnetic force streaming from the ends of the great earth magnet cause the Northern Lights, or *Aurora Borealis*.

Lodestone is a form of iron with a special chemical composition, and it is a natural magnet. Most interesting stories are told of the way the ancients discovered this apparently bewitched material, because it clung to the iron ends of their staffs or pulled the iron nails from their shoes. In the Ward's collection of minerals sent out to schools, which costs only one dollar, there is included a piece of lodestone, which is of perennial interest to the children.

Magnets made from lodestone are called natural magnets. A bar magnet or a horseshoe magnet has received its magnetism from some other magnet or from electrical sources. An electro-magnet is of soft iron, and is only a magnet when under the influence of a coil of wire charged with electricity. As soon as the current is shut off the iron immediately ceases to be a magnet.

LESSON

Leading thought— Any substance that will attract iron is called a magnet, and the force which enables it to attract iron is called magnetism. This force resides chiefly at the ends of magnets, called the poles. The forces residing at the opposite ends of a magnet act in opposite directions; in two magnets the like poles repel and the unlike poles attract each other. The needle of the mariner's compass points north and south, because the earth is a great magnet which has its south pole as a magnet at the North Pole of the world.

Method— Cheap toy horseshoe magnets are sufficiently good for this lesson, but the teacher should have a bar magnet, also a cheap toy compass, and a specimen of lodestone, which can be procured from any dealer in minerals. In addition, there should be nails, iron filings and tacks of both iron and brass, pins, darning needles or knitting needles, pens, etc. Each child, during play time, should have a chance to test the action of the magnets on these objects, and thus be able to answer for himself the questions which should be given a few at a time.

Observations—

1. How do we know that an object is a magnet? How many kinds of magnets do you know? Of what substance are the objects made which the magnets can pick up? Does a magnet pick up as many iron filings at its middle as at its ends? What does this show?

2. How far away from a needle must one end of the magnet be before the needle leaps toward it? Does it make any difference in this respect, if the magnet approaches the needle toward the point or

A horseshoe magnet

along its length? Does this show that the magnetic force extends out beyond the magnet? Does it show that the magnetic force works more strongly where it has more surface to act upon?

3. Take a tack and see if it will pick up iron filings or another tack. Place a tack on one end of the magnet, does it pick up iron filings now? What do you think is the reason for this difference in the powers of the tack?

4. Are some magnets stronger than others? Will some magnets pull the iron filings off from others? In the game of "stealing tacks," which can be played with two magnets, does each end of the magnet work equally well in pulling the tacks away from the other magnet?

5. Pick up a tack with a magnet. Hang another tack to this one end to end. How many tacks will it thus hold? Can you hang more tacks to some magnets than to others? Will the last tack picked up attract iron filings as strongly as the first next to the magnet? Why? Pull off the tack which is next to the magnet. Do the other tacks continue to hold together? Why? Instead of placing the tacks end to end, pick up one tack with the magnet and place others around it. Will it hold more tacks in this way? Why? If a magnet is covered with iron filings will it hold as many tacks without dropping the filings?

6. Take two horseshoe magnets and bring their ends together. Then turn one over and again bring the ends together. Will they cling to each other more or less strongly than before? Bring two ends of two bar magnets together; do they hold fast to each other? Change ends with one, now do the two magnets cling more or less closely than before? Does this show that the force in the two ends of a magnet is different in character?

7. Magnetize a knitting needle or a long sewing needle by rubbing one end of a magnet along its length twelve times, always in the same direction, *and not back and forth.* Does a needle thus treated pick up iron filings? Why?

8. Suspend this magnetized needle by a thread from some object where it can swing clear. When it finally rests does it point north and south or east and west?

9. Bring one end of a bar magnet or of a horseshoe magnet near to the north end of the suspended needle; what happens? Bring the other end of the magnet near the north end of the needle; what happens?

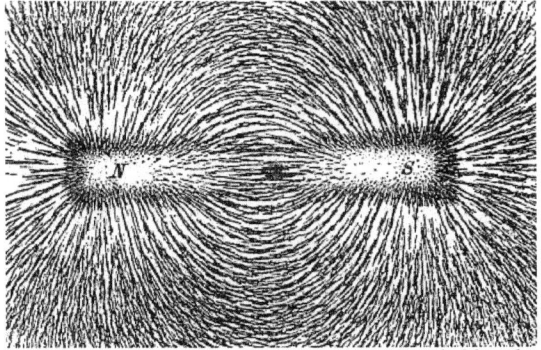
Iron filings on paper show the magnetic fields of a bar magnet

10. Magnetize two needles so that their eyes point in the same direction when they are suspended. Then bring the point of one of these needles toward the eye of the other, what happens? Bring the eye of one toward the eye of the other, what happens? When a needle is thus magnetized the end which turns toward the north is called the north pole, and the end pointing south is called the south pole.

11. Try this same experiment by thrusting the needles through the top of a cork and float them on a pan of water. Do the north poles of these needles attract or repel each other? Do the south poles of these needles attract or repel each other? If you place the north pole of one needle at the south pole of the other do they join and make one long magnet pointing north and south?

12. Take a pocket compass; place the north end of one of the magnetized needles near the north arm of the compass needle; what happens? Place the south pole of the needle near the north arm of the compass needle, what happens? Can you tell by the action of your magnet upon the compass needle which end of your magnet is the north pole and which the south pole?

13. Magnetize several long sewing needles by rubbing some of them toward the eye with the magnet and some from the eye toward the point. Take some small corks, cut them in cross sections about one-fourth inch thick, thrust a needle down through the center of each leaving only the eye above the cork. Then set them afloat on a

pan of water. How do they act toward each other? Try them with a bar magnet first with one end and then with the other, how do they act?

14. Describe how the needle in the mariner's compass is used in navigation.

15. Place fine iron filings on a pane of glass or on a stiff paper. Pass a magnet underneath; what forms do the filings assume? Do they make a picture of the direction of the lines of force which come from the magnet? Describe or sketch the direction of these lines of force, when the poles of a horseshoe magnet are placed below the filings. Place two similar poles of a bar magnet beneath the filings; what form do they take now?

16. What is lodestone? Why is it so called?

17. What is the difference between lodestone and a bar magnet? What is an electromagnet?

18. Write an English theme on "The Discovery and Early Use of the Mariner's Compass."

Supplementary reading— Electrical Experiments, Bonney; *The Wonder Book of Magnetism*, Houston; "The Third Royal Calendar" from *Arabian Nights Entertainments*.

"*Now, chief of all, the magnet's power I sing,*
And from what laws the attractive functions spring;
The magnet's name the observing Grecians drew
From the magnetic regions where it grew;
Its viewless potent virtues men surprise,
Its strange effects they view with wondering eyes,
When, without aid of hinges, links, or springs,
A pendant chain we hold of steely rings
Dropt from the stone—the stone the binding source,—
Ring cleaves to ring, and owes magnetic force:
Those held superior, those below maintain,
Circle 'neath circle downward draws in vain,
Whilst free in air disports the oscillating chain."

—"De Rerum Naturae," Lucretius, 93-52 B. C.

The Soil

TEACHER'S STORY

The soil is the sepulcher and the resurrection of all life in the past. The greater the sepulcher the greater the resurrection. The greater the resurrection the greater the growth. The life of yesterday seeks the earth to-day that new life may come from it tomorrow. The soil is composed of stone flour and organic matter (humus) mixed; the greater the store of organic matter the greater the fertility.

—JOHN WALTON SPENCER.

BECAUSE the child, after making mud pies, is told that his face is dirty, he naturally concludes that all soil is dirt. But it is only when out of place that it is dirt; for, in place, it is the home of miracles—the matrix from which comes that wonderful force which we call life. After the study of the brook, the crystals, the minerals and the rocks, the pupils are ready for a more careful study of the soil. However, most of the study in soils belongs to agriculture rather than to nature-study.

The brook mill even at low water grinds ceaselessly, sorting out the finer products and carrying them away to serve as soil material

THE SOIL MAKERS

If we could go back to the very beginning, we should find that the soil consisted solely of broken off particles of rock—particles so finely ground by nature's forces that we might properly call them "rock flour." In our study of the brook, we noted that those stones with sharp corners were just beginning their experience in the brook mill, and those that were rounded out, forming pebbles, had their corners ground off in the making of the soil grist. And in the work of the brook we saw how this grinding was done, and how the soil grist is sifted, sorted, carried and dropped.

But there are other agencies besides water that help in grinding the stone flour. If we visit some rocky cliff, we are sure to find at its base a heap of stones, gravel and soil, which the geologists call *talus*. In our eastern country we know that these pebbles and soil were pried loose by Jack Frost with his ice wedges. The water filters into all the cracks and crevices of the rock, and since water, when freezing, is obliged to expand, the particles of rock were thereby torn loose and forced off and fell to the bottom of the cliff. Moreover, rocks expand when hot, and are often thus broken without the aid of water and frost. In the rocks of the desert, the changes in temperature pry off the rock particles, which the winds carry away to make up the sands of the desert. The winds hurl these sands against other rocks which are still standing, and hurl them with such force that more particles are torn off, making more sand. In fact, the wind, in some regions,

Lichen growing on rocks

grinds the rocks into stone flour as effectually as does the water in other places. Then, too, the gases of the air also cause rocks to decay. We know how iron rusts and falls to pieces through contact with the gases of the air. Some rocks decompose in a similar way. We often see that the inscriptions on old headstones have been almost obliterated, because the gases in the air have so decomposed the marble.

In addition to the other soil makers, there are the little plants which we call lichens. The spores of these plants are so minute that we cannot see them, and they drift about in the air until they find resting place upon some rock. Here they begin to grow, and as they grow they become strongly acid; they are thus enabled to eat a foothold into the rock, softening its surface and powdering it into stone flour. And in these situations other plants grow later, sending their roots down into every crack and crevice and thus prying off more of the rock.

THE SOIL CARRIERS

In the study of the brook we have seen how the water lifts, carries and deposits the soils; and since, at one time or another, the entire surface of the earth has been under water, we can see that water has

Boulders left by a glacier

been the most important of the soil carriers and has done the greatest work. The wind carries much soil, especially in the arid regions; the movements of the sand dunes in the deserts and on the seashores bear witness to what the wind can do as a soil carrier. But in the northern United States, from New England to the Dakotas, much of our soil has been carried by a great ice river that once upon a time flowed down upon our lands from the North. This great, slow-moving river, perhaps a mile or more high, plowed up the soil and stones, and freezing them fast carried and shoved them along under its great weight. After a time the ice melted and dropped its burden. Many of the stones were of granite taken up from the old mountains of northern Canada and ground off and rounded during their journey. We call these stones which were brought down to us from the North, "boulders;" and the soils which were brought along on the bottoms of glaciers and dropped and pressed down by the tremendous ice weight and thus made compact although unsorted, we call "hardpan."

THE KINDS OF SOIL

By the work of these soil makers and soil carriers, the rock flour was made. But if we should take some of it and plant our seeds in it, we should find that they would not grow thriftily, even though we watered them and gave them every care. The reason for this is that most rock flour does not have in it the substances which the plants most need for their growth. But if we should go to the woods and get some of the black woods-earth and mix it with rock flour, we should find that our plants would thrive. This rich, earth mold in the forest is almost wholly made up of matter once alive, but which is now decayed, and which we call "humus." The more humus that we have in the rock flour, the richer it is in plant food, and the more plant growth it will support.

In general, soils may be divided into clay, sand, gravel, loam and humus.

Clay in its purest state is kaolinite, the result of weathering of feldspar, or mica. It is finely powdered

Soil layers

and is used for pottery, while the less pure clays are used for brick-making. Clayey soil is sticky and slippery when wet, and bakes hard and cracks when dry. It is hard to cultivate, but it absorbs moisture from the air and holds fast to its fertility, and is especially good for permanent pastures and meadows.

Sand, in a pure state, is made up mostly of finely broken particles of quartz and feldspar, and is used for the making of glass. A sandy soil is light and open and easy to work. It absorbs little water from the air and has little power for holding plant food, since the water washes it out. It is especially valuable for truck gardening, because it is a warm soil. It is warm because water does not evaporate from its surface rapidly.

Humus is composed of decayed animal and vegetable matter. It is very rich in plant food. Wherever there is humus in the soil it is likely to be darker in color than the stone flour.

Loam is a mixture of clay, sand and humus. For many crops it is the most desirable soil.

LESSON

Leading thought— The soil is composed of rock flour and humus. Soil, to support life, must be porous, so that the roots of the plants may receive through it both water and air.

Method— The children should bring in specimens of soils from various localities near the school. Parts of each specimen should be wet to see if they are clayey, that quality showing quickly in the putty-like adhesiveness when rubbed between the fingers. It would be well to get some pure blue clay, and let the children make marbles of it to impress upon them this quality of clay. They should try and make marbles of other soils to show the lack of adhesiveness in them. They should examine sand through a lens and should examine humus in a similar way. After they are familiar with these three kinds of soils, they are ready for the lesson.

Rich black loam ready to grow vegetables

Observations—

1. Look at any kind of soil with a lens, and tell why you think it is made up of small pieces of stone and rock.

2. Take a piece of rock and pound it fine. What does it look like? Do you think that your plants will grow well if you plant them in the rock flour which you have just made? Try the experiment and describe the results.

3. How does the water grind off the stones and make soil? How does the wind do it?

4. How do water and frost pry off pieces of rock? Is there a cliff in your neighborhood that has at its foot a heap of soil and stones? Where did these comes from?

5. How do the lichens and other plants pry off the outside of rocks? Have you ever found lichens growing on stones?

6. Have you ever noticed old headstones in the cemetery that were falling to pieces? What causes them to decay?

7. Write an English theme on the great glacier that formerly covered the northeastern portion of the United States.

8. Go to the woods, scrape off the leaves and get some of the black earth beneath them. Of what is this soil composed? Is it rock flour?

Loam. Sand. Clay.
Note the sand has allowed the most water to drop through it,
the loam next, while no water has passed through the clay

What makes it so black? Why do you call this soil rich? What does it do if you add it to the soil in the pots where your flowers are growing?

9. Find a railroad cut or some other place where the earth is exposed for some distance up and down. Is there solid rock at the bottom? How deep is the soil above the rock? Is the soil the same color at the surface as it is below? Why is this?

10. *Experiment 1: To show which kinds of soil hold most water*— Take three lamp chimneys, or bottles from which the bottoms have been broken. Place in one loam, in another clay, in another fine-grained sand, using in each case the same amount. Tie cheesecloth over the bottom, so that the soil will not fall out; make the soil compact by jarring down. Place each over a tumbler. From a cup of water, held as near as possible to the soil, pour water into one of the bottles slowly, so as to keep the surface of the soil covered. Consult a watch and note how long before the water begins dripping below. Do the same with the other two. Compare the results. Which soil takes the water most rapidly? Which lets it through first? Which lets through the most? How would rain affect fields of clayey soil? Of sandy soil? Of loam?

Hints for teacher on Experiment No. 1— Through sand the water passes very rapidly—in less than a minute if the sand is coarse. It takes several minutes (14 min.) to go through loam, but requires some hours to appear below the clay. It requires more water to saturate clay.

Sand. Clay. Loam.
The water has nearly reached the upper surface of the sand and is halfway up the loam; in the clay it has climbed but a short distance.

Care should be taken to use the same amount of water on the three kinds of soil. More than one application will be required for clay, since the amount of water accommodated in the chimney above the soil will not be sufficient to saturate clay.

More water will be found to have percolated through sand than through loam or clay. The latter are more retentive of moisture than is sand, although absorbing rain less readily than sand. The mixture of sand and clay in loam is most ideal for cultivated fields, absorbing moisture more readily than clay and retaining it better than sand.

The unmulched loam in the chimney at the left dried out in four days. The loam covered with a dust mulch in the other chimney retained moisture for a month.

Experiment 2— Fill a glass tumbler with very small marbles or buckshot. Pour water over them to fill the glass. Placing cheesecloth over the top of the tumbler pour off all the water that easily drains away. Remove the cheesecloth, and immediately examine the marbles for the film of water which surrounds each one and can clearly be seen where one marble comes in contact with another marble or the side of the glass.

Hints for teacher on Experiment 2— It is such a film of water as remains on the marbles that on each particle of soil supplies the plant with water and food. The water between the marbles has been drained off. This water corresponds to that carried out of the soil by drainage; it is injurious to the plant, keeping "its feet too wet," and should be removed.

Experiment 3—To show that soil lifts water up from below— Use the same soils arranged in the same way as for Experiment 1, but instead of pouring water in at the top, place the three lamp chimneys in a pan which has water in it about an inch deep. In which soil does the water rise most rapidly? In which does it rise the highest? After the water has been taken up, let the soil stand in the lamp chimneys for several days. Which soil dries out the soonest? If we had three fields, one of loam, one of clay, and one of sand, in which would the most water be lifted from below for the use of the plants? Which would retain the water longest?

Hints for teacher on Experiment 3— Water rises through the sand in a short time; if rather fine sand is used it requires less than half an hour. To rise through loam it will require three or four times as long, and may not reach the top of the clay for several days. If the glass tubes were three or four feet long and allowed to stand for several days, we would find that although the water climbs very slowly through the clay it will climb to a greater height in clay than in loam or sand. Under field conditions clay will retain moisture for a longer time than sand or loam.

Experiment 4—To show that mulch keeps the water from evaporating from soils— Take two of the lamp chimneys filled half full with loam. Pour in the same amount of water in each until the soil is thoroughly wet. Cover the top of one with an inch deep of dry, loose earth. Which dries out first? What does the loosening and pulverizing of the soil in our fields by harrowing do for our planted crops? What is a mulch?

Hints for teacher on Experiment 4— The soil covered with a layer of dry soil—a dust mulch—will retain moisture much longer than the unmulched soil. Hence, the farmer or gardener loosens and pulverizes the top soil by harrowing, hoeing or raking in order to retain moisture for plant roots. A mulch may also be of straw, boards, leaves or stones. Fallen leaves form a natural mulch in the woods. There, at any time, under this covering, may be found moist earth. A mulch is a

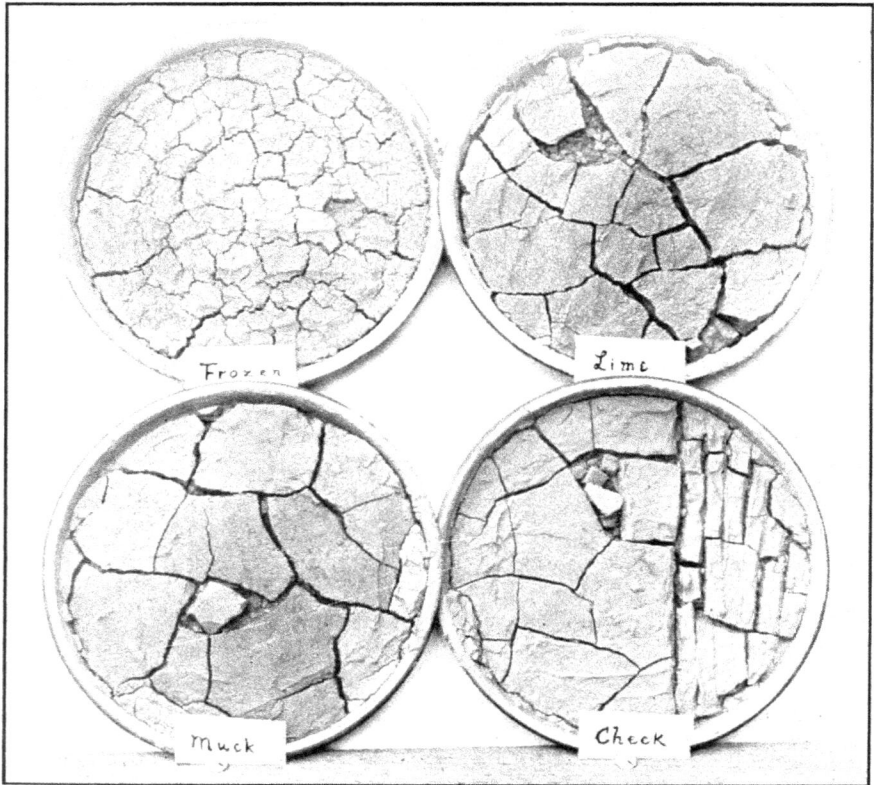

Experiment to show the proper treatment of clay soil.

soil cover which breaks the capillary connection, so that water will not rise to the surface to be evaporated. To be efficient a mulch must be *dry*. After rain the "dust blanket" on the garden bed should be renewed by cultivation.

Experiment 5— Fill several vials with different soils from fields in the neighborhood. If the soil in any of the vials is dry, moisten it. Take a piece of blue litmus paper and press down into the soil in each vial. Does the litmus paper turn red as it becomes dampened by the soil in any of the vials? If so, this soil is acid. Add a little lime and mix it in thoroughly with the soil in the vial that shows the acid soil. Test it again with the litmus paper. Does the paper remain blue or turn red? Does alfalfa and clover grow on acid soils? Why should we add lime to such soils?

Hints for teacher on Experiment 5— A slightly acid soil may show no reaction with litmus paper. It may be well to have a prepared soil with a few drops of vinegar or other acid added, which will show the reaction. The addition of lime will correct the acid condition. Soils for alfalfa or clover should never be acid. They are usually well limed before an attempt is made to grow these legumes.

Experiment 6, which indicates the proper treatment of clay soils— Fill four pie tins with clay which has been wet and smoothly puddled. In one mix with the clay a small portion of lime; in another add a larger portion of muck; leave two with pure clay, and place one of these out-of-doors where it will freeze hard. Then place the four tins on a shelf and allow to dry. In which of these is the clay most friable? In which is it the hardest?

Hints to the teacher on Experiment 6— This experiment shows that freezing the clay rendered it finer, so that it may be broken easily into particles small enough to set closely about the plant's roots. The clay mixed with lime is much more friable than the one mixed with muck, showing that clay needs lime more than organic matter to make it of greatest use. The pure clay which is dried without freezing hardens into large, flat pieces, each being almost as hard as stone.

Supplementary reading— Ch. I, II, III in *The Great World's Farm*, Gaye: Ch. IV. in *Practical Forestry*, Gifford.

Beside the moist clods the slender flags arise filled with the sweetness of the earth. Out of the darkness—under that darkness which knows no day save when the ploughshare opens its chinks—they have come to the light. To the light they have brought a colour which will attract the sunbeams from now till harvest.

—Richard Jefferies.

"Here is a problem, a wonder for all to see.
Look at this marvelous thing I hold in my hand!
This is a magic surprising, a mystery
Strange as a miracle, harder to understand.
What is it? Only a handful of dust: to your touch

A dry, rough powder you trample beneath your feet,
Dark and lifeless; but think for a moment, how much
It hides and holds that is beautiful, bitter, or sweet.
Think of the glory of color! The red of the rose,
Green of the myriad leaves and the fields of grass,
Yellow as bright as the sun where the daffodil blows,
Purple where violets nod as the breezes pass.
Strange, that this lifeless thing gives vine, flower, tree,
Color and shape and character, fragrance too;
That the timber that builds the house, the ship for the sea,
Out of this powder its strength and its toughness drew!"

—FROM "DUST," CELIA THAXTER.

Some years ago there was received at Cornell University a letter from a boy working upon a farm in Canada. In this letter he said:

"I have read your leaflet entitled, 'The Soil, What It Is,' and as I trudged up and down the furrows every stone, every lump of earth, every shady knoll, every sod hollow had for me a new interest. The day passed, the work was done, and I at least had had a rich experience."

Fog on the mountains

Water Forms

TEACHER'S STORY

WATER, in its various changing forms, is an example of another overworked miracle—so common that we fail to see the miraculous in it. We cultivate the imagination of our children by tales of the Prince who became invisible when he put on his cap of darkness, and who made far journeys through the air on his magic carpet. And yet no cap of darkness ever wrought more astonishing disappearances than occur when this most common of our earth's elements disappears from under our very eyes, dissolving into thin air. We cloak the miracle by saying "water evaporates," but think once of the travels of one of these drops of water in its invisible cap! It may be a drop caught and clogged in a towel hung on the line after washing, but as soon as it dons its magic cap, it flies off in the atmosphere invisible to our eyes; and the next time any of its parts are evident to our senses, they may occur as a portion of the white masses of cloud sailing across the blue sky, the cloud which Shelley impersonates:

"I am the daughter of Earth and Water,
And the nursling of the Sky;
I pass through the pores of the ocean and shores;
I change, but I cannot die."

We have, however, learned the mysterious key-word which brings back the vapor spirit to our sight and touch. This word is "cold." For if our drop of water, in its cap of darkness, meets in its travels an object which is cold, straightway the cap falls off and it becomes visible. If it be a stratum of cold air that meets the invisible wanderer, it becomes visible as a cloud, or as mist, or as rain. If the cold object be an ice pitcher, then it appears as drops on its surface, captured from the air and chained as "flowing tears" upon its cold surface. And again, if it be the cooling surface of the earth at night that captures the wanderer, it appears as dew.

But the story of the water magic is only half told. The cold brings back the invisible water vapor, forming it into visible drops; but if it is cold enough to freeze, then we behold another miracle, for the drops are changed to crystals. The cool window-pane at evening may be dimmed with mist caught from the air of the room; if we examine the mist with a lens we find it composed of tiny drops of water. But if the night be very cold, we find next morning upon the window-pane exquisite ferns, or stars, or trees, all formed of the crystals grown from the mist which was there the night before. Moreover, the drops of mist have been drawn together by crystal magic, leaving portions of the glass dry and clear.

If we examine the grass during a cool evening of October we find it pearled with dew, wrung from the atmosphere by the permeating coolness of the surface of the ground. If the following night be freezing cold, the next morning we find the grass blades covered with the beautiful crystals of hoar frost.

If a raincloud encounters a stratum of air cold enough to freeze, then what would have been rain or mist comes down to us as sleet, hail or snowflakes, and of all the forms of water crystals, that of snow in its perfection is the most beautiful; it is, indeed, the most beautiful of all crystals that we know. Why should water freezing freely in the air so demonstrate geometry by forming, as it does, a star with six rays, each set to another, at an angle of 60 degrees? And as if to prove geometry divine beyond cavil, sometimes the rays are only three in number—a factor of six—and include angles of twice 60 degrees. Moreover, the rays are decorated, making thousands of intricate and beautiful forms; but if one ray of the six is ornamented with additional crystals the other five are decorated likewise. Those

Dew on a spider's web

snow crystals formed in the higher clouds and, therefore, in cooler regions may be more solid in form, the spaces in the angles being built out to the tips of the rays including air spaces set in symmetrical patterns: and some of the crystals may be columnar in form, the column being six-sided. While those snow crystals formed in the lower currents of air, and therefore in warmer regions, show their six rays marvellously ornamented. The reason why the snow crystals are so much more beautiful and perfect than the crystals of hoar frost or ice, is because they are formed from water vapor, and grow freely in the regions of the upper air. Mr. W. A. Bentley, who has spent many years photographing the snow crystals, has found more than 1300 distinct types.

The high clouds are composed of ice crystals formed from the cloud mists; such ice clouds form a halo when veiling the sun or the moon.

When the water changes to vapor and is absorbed into the atmosphere, we call the process evaporation. The water left in an open saucer will evaporate more rapidly than that in a covered saucer, because it comes in contact with more air. The clothes which are hung on the line wet, dry more rapidly if the air is dry and not damp; for if the air is damp, it means that it already has almost as much water in it as it can hold.

Composite snow crystal formed in high and medium clouds

Frost on a fence

The clothes will dry more rapidly when the air is hot, because hot air takes up moisture more readily and holds more of it than does cold air. The clothes will dry more rapidly on a windy day, because more air moves over them and comes in contact with them than on a still day.

If we observe a boiling teakettle, we can see a clear space of perhaps an inch or less in front of the spout. This space is filled with steam, which is hot air saturated with hot water vapor. But what we call "steam" from a kettle, is this same water vapor condensed back into thin drops of water or mist by coming into contact with the cooler air of the room. When the atmosphere is dry, water will boil away much more rapidly than when the air is damp.

The breath of a horse, or our own breath, is invisible during a warm day; but during a cold day, it is condensed to mist as soon as it is expelled from the nostrils and comes in contact with the cold air. The one who wears spectacles finds them unclouded during warm days; but in winter the glasses become cold out of doors, and as soon as they are brought into contact with the warmer, damp atmosphere of a room, they are covered with a mist. In a like manner, the window-pane in winter, cooled by the outside temperature, condenses on its inner surface the mist from the damp air of the room.

The water vapor in the atmosphere is invisible, and it moves with the air currents until it is wrung out by coming into contact with the cold. The air thus filled with water vapor may be entirely clear near the surface of the earth; but, as it rises, it comes in contact with cooler air and discharges its vapor in the form of mist, which we call clouds; and if there is enough vapor in the air when it meets a cold current, it is discharged as rain and falls back to the earth. Thus, when it is very cloudy, we think it will rain, because clouds consist of mist or fog; and if they are subjected to a colder temperature, the mist is condensed to rain. Thus, often in mountainous regions, the fog may be seen streaming and boiling over a mountain peak, and yet always disappears at a certain distance below it. This is because the temperature around the peak is cold and condenses the water vapor as fast as the wind brings it along, but the mist passes over and soon meets a warm current below and, presto, it disappears! It is then taken back into the atmosphere. The level base of a cumulus cloud has a stratum of warmer air below it, and marks the level of condensation.

At the end of the day, the surface of the ground cools more quickly than the air above it. If it becomes sufficiently cold and the air is damp, then the water from it is condensed and dew is formed during the night. However, all dew is not always condensed from the atmo-

Ice taking over a lake

Many types of cloud over the mountains. Note dark rain clouds on the right

sphere, since some of it is moisture pumped up by the plants, which could not evaporate in the cold night air. On windy nights, the stratum of air cooled by the surface of the earth is moved along and more air takes its place, and it therefore does not become cold enough to be obliged to yield up its water vapor as dew. If the weather during a dewy night becomes very cold, the dew becomes crystallized into hoar frost. The crystals of hoar frost are often very beautiful and are well worth our study.

The ice on the surface of a still pond begins to form usually around the edges first, and fine, lancelike needles of ice are sent out across the surface. It is a very interesting experience to watch the ice crystals form on a shallow pond of water. This may easily be seen during cold winter weather. It is equally interesting to watch the formation of the ice crystals in a glass bottle or jar. Water, in crystallizing, expands, and requires more room than it does as a fluid; therefore, as the water changes to ice it must have more room, and often presses so hard against the sides of the bottle as to break it. The ice in the surface soil of the wheat fields expands and buckles, holding fast in its grip the leaves of the young wheat and tearing them loose from their roots; this "heaving" is one cause for the winter-killing of wheat. Sleet consists of rain crystallized in the form of sharp needles. Hail consists of ice and snow compacted together, making the hard, more or less globular hailstones.

LESSON

Leading thought— Water occurs as an invisible vapor in the air and also as mist and rain; and when subjected to freezing, it crystallizes into ice and frost and snow.

Method— The answers to the questions of this lesson should, as far as possible, be given in the form of a demonstration. All of the experiments suggested should be tried, and the pupils should think the matter out for themselves. In the study of the snow crystals a compound microscope is a great help, but a hand lens will do. This part of the work must be done out of doors. The most advantageous time for studying the perfect snow crystals is when the snow is falling in small, hard flakes; since, when the snow is soft, there are many crystals massed together into great fleecy flakes, and they have lost their original form. The lessons on frost or dew may be given best in the autumn or spring.

Observations—

1. Place a saucer filled with water near a stove or radiator; do not cover it nor disturb it. Place another saucer filled with water near this but cover it with a tight box. From which saucer does the water evaporate most rapidly? Why?

2. We hang the clothes, after they are washed, out of doors to dry; what becomes of the water that was in them? Will they dry

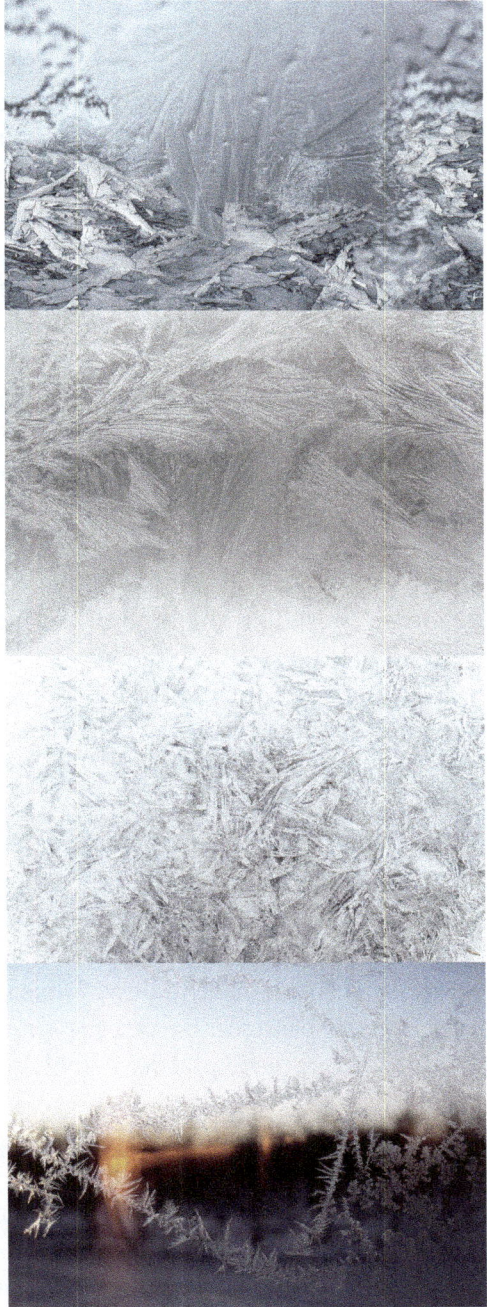

Different front patterns on a window

more rapidly during a clear or during a damp day? Why? Will they dry more rapidly during a still or during a windy day? Why? Will they dry more rapidly during hot or cold weather? Why?

3. Watch a teakettle of water as it is boiling. Notice that near its spout there is no mist, but what we call steam is formed beyond this. Why is this so? What is steam? Why does water boil away? Do kettles boil dry sooner on some days than on others? Why?

4. If the water disappears in the atmosphere where does it go? Why do we say "the weather is damp"? What force is it that wrings the water out of the atmosphere?

5. Why does the breath of a horse show as a mist on a cold day? Why do persons who wear spectacles find their glasses covered with mist as soon as they enter a warm room after having been out in the cold? Why do the window-panes become covered with mist during cold weather? Is the mist on the outside or on the inside? Why does steam show as a white mist? Why does the ice pitcher, on a warm day, become covered on the outside with drops of water? Would this happen on a cold day? Why not?

6. Why, when the water is invisible in the atmosphere, does it become visible as clouds? What causes the lower edges of cumulus clouds to be so level? What is fog? Why do clouds occur on mountain peaks? What causes rain?

7. What causes dew to form? When the grass is covered with dew, are the leaves of the higher trees likewise covered? Why not? What kind of weather must we have in order to have dewy nights? What must be the atmosphere of the air in relation to that of the ground in order to condense the dew? Does dew form on windy nights? Why not? Does all dew come from the air, or does some of it come from the ground through the plants? Why is not this water, pumped up by the plants, evaporated?

8. What happens to the dew if the weather becomes freezing during the night? What is hoar frost? Why should water change form when it is frozen? How many forms of frost crystals can you find on the grass on a frosty morning?

9. When a pond begins freezing over, what part of it freezes first? Describe how the first layer of ice is formed over the surface.

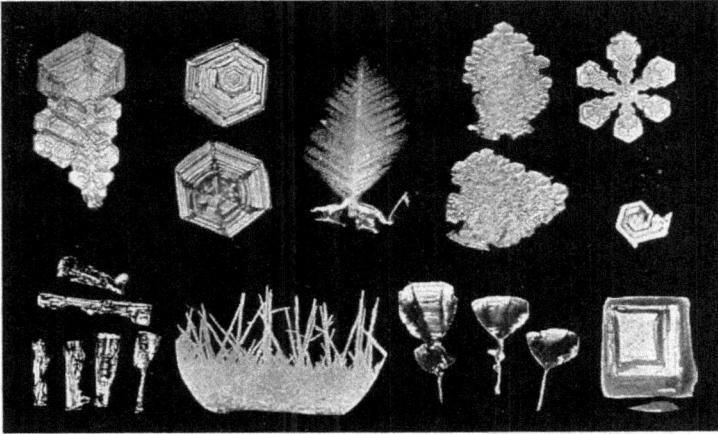

Forms of hoar frost

10. Place a bottle of water out of doors in freezing weather. How does the ice appear in it at first? What happens later? Why does the bottle break? How is it that water which has filled the crevices of rocks scales off pieces of the rock in cold weather? Why does winter wheat "winter-kill" on wet soil?

11. Why does frost form on a window-pane? How many different figures can you trace on a frosted pane? Are there any long, needlelike forms? Are there star forms? Can you find forms that resemble ferns and trees? Do you sometimes see, on boards or on the pavement, frost in forms like those on the window-pane?

12. When there is a fine, dry snow falling, take a piece of dark flannel and catch some flakes upon it. Examine them with a lens, being careful not to breathe upon them. How many forms of snow crystals can you find? How many rays are there in the star-shaped snow crystals? Do you find any solid crystals? Can you find any crystals that are triangular? When the snow is falling in large, feathery flakes, can you find the crystals? Why not?

13. What is the difference between a hailstone and a snow crystal? What is sleet?

Supplementary reading— Water Wonders, Thompson; *Forms of Water*, Tyndall.

"When in the night we wake and hear the rain
Which on the white bloom of the orchard falls,
And on the young, green wheat-blades,
where thought recalls
How in the furrow stands the rusting plow,
Then fancy pictures what the day will see—
The ducklings paddling in the puddled lane,
Sheep grazing slowly up the emerald slope,
Clear bird-notes ringing, and the droning bee
Among the lilac's bloom—enchanting hope—
How fair the fading dreams we entertain,
When in the night we wake and hear the rain!"

—ROBERT BURNS WILSON.

"The thin snow now driving from the north and lodging on my coat consists of those beautiful star crystals, not cottony and chubby spokes, but thin and partly transparent crystals. They are about a tenth of an inch in diameter, perfect little wheels with six spokes without a tire, or rather with six perfect little leaflets, fern-like, with a distinct straight and slender midrib,

Snowflakes

raying from the center. On each side of each midrib there is a transparent thin blade with a crenate edge. How full of creative genius is the air in which these are generated! I should hardly admire more if real stars fell and lodged on my coat. Nature is full of genius, full of divinity. Nothing is cheap and coarse, neither dewdrops nor snowflakes."

"A divinity must have stirred within them before the crystals did thus shoot and set. Wheels of storm-chariots. The same law that shapes the earth-star shapes the snow-stars. As surely as the petals of a flower are fixed, each of these countless snow-stars comes whirling to earth, pronouncing thus, with emphasis, the number six."

—THOREAU'S JOURNAL.

The Weather

—Wilford M. Wilson
Section Director, U. S. Weather Bureau,
and Professor of Meteorology in Cornell University.

THE atmosphere, at the bottom of which we live, may be compared to a great ocean of air, about two hundred miles deep, resting upon the earth. The changes and movements that take place in this ocean of air, the storms that invade it, the clouds that float in it, the sunshine, the rain, the dew, the sleet, the frost, the snow, and the hail are termed "weather." We live in it; we partake of its moods; we reflect its sunshine and shadows; it invades the everyday affairs of life, influences every business and social activity, and molds the character of nations; and yet nearly everything we know about the weather has been learned within the lifetime of the present generation. Not that the weather did not interest men of early times, but the problem appeared to be so complicated and so complex that it baffled their utmost endeavors.

The Tower of the Winds showing the wind gods Boreas (north wind on the left) and Skiron (northwesterly wind, on the right)

THE TEMPLE OF THE WINDS AT ATHENS

The Temple of the Winds, erected probably about five hundred years B. C., indicates the knowledge of the weather possessed by the ancient Greeks. This temple is a little octagon tower, the eight sides of which face the eight principal winds. On each of its eight sides is a human figure cut in the marble, symbolizing the kind of weather the wind from that particular direction brought to Athens.

Boreas, the cold north wind, is represented by the figure of an old man wearing a thick mantle, high buskins (boots) and blowing on a "weathered horn." The northeast wind, which brought, and still brings to Athens, cold, snow, sleet and hail, is symbolized by a man with a severe countenance who is rattling sling-stones in a shield, thus expressing the noise made by the falling hail and sleet.

The east wind, which brought weather favorable to the growth of vegetation, is shown by the figure of a beautiful youth bearing fruit and flowers in his tucked-up mantle.

Natos, the warm south wind, brought rain, and he is about to pour the water over the earth from the jar which he carries.

Lips, the southwest wind, beloved of the Greek sailors, drives a ship before him, while Zephros, the gentle west wind, is represented by a youth lightly clad, scattering flowers as he goes.

The northwest wind, which brought dry and sometimes hot weather to Athens, is symbolized in the figure of a man holding a vessel of charcoal in his hands. Thus, the character of the weather brought by each separate wind is fixed in stone, and from this record we learn that, even with the lapse of twenty centuries, there has come no material change.

HISTORICAL

There is no record of any rational progress having been made in the study of the weather until about the middle of the seventeenth century, when Torricelli discovered the principles of the barometer. This was a most important discovery and marks the beginning of the modern science of meteorology. Soon after Torricelli's discovery of the barometer his great teacher, Galileo, discovered the thermometer, and thus made possible the collection of data upon which all meteorological investigations are based. About one hundred years after the discovery of the barometer, Benjamin Franklin made a discovery of equal importance. He demonstrated that storms were eddies in the atmosphere, and that they progressed or moved as a whole, along the surface of the earth.

It might be interesting to learn how Franklin made this discovery. Franklin, being interested at that time in astronomy, had arranged with a friend in Boston to take observations of a lunar eclipse at the same time that he, himself, was to take observations at Philadelphia. On the night of the eclipse a terrific northeast wind and rain storm set in at Philadelphia, and Franklin was unable to make any observations. He reasoned, that as the wind blew from the northeast, the storm must have been experienced in Boston before it reached Philadelphia. But imagine his surprise, when he heard from his friend in Boston that the night had been clear and favorable for observation, but that a fierce wind and rain storm set in on the following morning. Franklin determined to investigate. He sent out letters of inquiry

Lightning crawling across the sky

to all surrounding mail stations, asking for the time of the beginning and ending of the storm, the direction and strength of the wind, etc. When the information contained in the replies was charted on a map it showed that, at all places to the southwest of Philadelphia, the beginning of the storm was earlier than at Philadelphia, while at all places to the northeast of Philadelphia the beginning of the storm was later than at Philadelphia. Likewise, the ending was earlier to the southwest and later to the northeast of Philadelphia than at Philadelphia. He also found that the winds in every instance passed through a regular sequence, setting in from some easterly point and veering to the south as the storm progressed, then to the southeast and finally to the west or northwest as the storm passed away and the weather cleared.

A further study of these facts convinced Franklin that the storm was an eddy in the atmosphere, and that the eddy moved as a whole from the southwest toward the northeast, and that the winds blew from all directions toward the center of the eddy, impelled by what he termed suction.

Franklin was so far in advance of his time that his ideas about storms made little impression on his contemporaries, and so it remained for Redfield, Espy, Loomis, Henry and Maury and other American meteorologists, a hundred years later, to show that Franklin had gained the first essentially correct and adequate conception of the structure and movement of storms.

During the first half of the nineteenth century, considerable progress was made in the study of storms, principally by American meteorologists, among whom was William Redfield of New York, who first demonstrated that storms had both a rotary and progressive movement. James Espy followed Redfield in the construction of weather maps, although he had already published much on meteorological subjects before the latter entered the field.

Professor Joseph Henry, Secretary of the Smithsonian Institution at Washington, was the first to prepare a daily weather map from observations collected by telegraph. He made no attempt to make forecasts, but used his weather map to demonstrate to members of Congress the feasibility of a national weather service.

An incident occurred during the Crimean War that gave meteorology a great impetus, especially in Europe. On November 10th, 1854, while the French fleet was at anchor in the Black Sea, a storm of great intensity occurred which practically destroyed its effectiveness against the enemy. The investigation that followed showed that the storm came from western Europe, and had there been adequate means of communication and its character and direction of progress been known, it would have been possible to have warned the fleet of its approach and thus afforded an opportunity for its protection.

This report created a profound impression among scientific men and active measures were taken at once that resulted in the organization of weather services in the principal countries of Europe between 1855 and 1860.

The work of Professor Henry, Abbe, and others in this country would, doubtless, have resulted in such an organization in the United States in the early 60's, had not the Civil War intervened, absorbing public attention to the exclusion of other matters. It was not until 1870, that Dr. Increase A. Lapham of Milwaukee, in conjunction with Representative Paine of that city, was able so to present the claims for a national weather service that the act was finally passed that gave birth to the present meteorological bureau in the United States. Dr. Lapham issued from Chicago on November 10, 1871, the first official forecast of the weather made in this country.

THE ATMOSPHERE

What is known about the atmosphere of our earth has been learned from the exploration of a comparatively thin layer at the bottom. There is reason to believe that the atmosphere extends upwards about two hundred miles from the surface of the earth. We have a great mass of observations made at the surface, some on mountains, but few in the free air more than a few miles above the surface. Our knowledge of the upper atmosphere is, therefore, in the nature of conclusions drawn from such observations as are at hand, and is subject to changes and modifications as the facts become known by actual observation.

During the past few years a concerted effort has been made in various parts of the world to explore the upper atmosphere by means of kites and balloons, carrying meteorological instruments that automatically record the temperature, pressure, humidity, velocity and direction of the wind, etc. In this country this work has been carried on principally at the Mount Weather Observatory, which is located in Loudon County, Virginia, and is under the direction of the United States Weather Bureau and at Blue Hill Observatory, a private institution located near Boston and supported by Professor Lawrence Rotch. From observations thus obtained much has been learned about the upper atmosphere that was not even suspected before. Some theories have been confirmed and some destroyed, but this line of research is gradually bringing us nearer the truth.

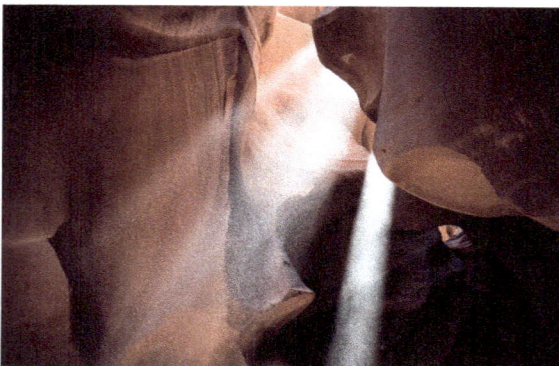

Dust floating in the air

AIR AS A GAS

Air is not a simple substance, as was once supposed, but is composed of a number of gases, each one of which tends to form an atmosphere of its own, just as it would if none of the other gases were present. The different gases of the atmosphere are not chemically combined but are very thoroughly mixed, as one might mix sugar and salt. Samples of air collected from all parts of the world show that the relative proportion of the gases forming the atmosphere is practically uniform.

THE COMPOSITION OF AIR

Dry air is composed chiefly of oxygen and nitrogen. There are, however, small quantities of carbon dioxide, argon, helium, krepton, neon, hydrogen and xenon, and probably other gases yet to be discovered.

The approximate proportion, by volume, is as follows: Nitrogen 78 parts, oxygen 21 parts, argon 1 part, carbon-dioxide .03 parts, and krepton, helium and xenon a trace. Pure dry air does not exist in nature, so there is always present in natural air a variable amount of water vapor, depending upon the temperature and the source of supply. Besides these, which may be termed the permanent constituents of the atmosphere, many other substances are occasionally met with. Lightning produces minute quantities of ammonia, nitrous acid and ozone. Dust comes from the earth, salt from the sea, while innumerable micro-organisms, most of which are harmless, besides the pollen and spores of plants, are frequently found floating in the atmosphere. Recent investigations in atmospheric electricity lead to the conclusion that electric ions are also present, and perform important functions, especially with respect to precipitation.

OXYGEN

Oxygen is one of the most common substances. It exists in the atmosphere as a transparent, odorless, tasteless gas. It combines with hydrogen to form the water of the oceans, and with various other substances to form much of the solid crust of the earth. Chemically, it is a very active gas, and because of its tendency to unite with other substances to form chemical compounds, it is believed that the volume of oxygen now in the atmosphere, is less than during the early history of the earth. It supports combustion by combining with carbon and other substances, producing light and heat. It combines with some of the organic constituents of the blood, through the function of respiration, which is in itself a slow process of combustion, and thus supports life and maintains the bodily heat.

NITROGEN

Nitrogen forms the largest proportion of the atmosphere, but unlike oxygen it is a very inert substance, uniting with no element at ordinary temperatures, and at high temperatures with only a few; and when so united the bonds that hold it are easily broken and the gas set free. For this reason, it is utilized in the manufacture of explosives, such as gunpowder, guncotton, nitroglycerine, dynamite, etc. Its office in the atmosphere appears to be to give the air greater weight and to dilute the oxygen, for in an atmosphere of pure oxygen a fire once started could not be controlled. Although nitrogen does not contribute directly to animal life, in that it is not absorbed and assimilated from the air direct as oxygen is, nevertheless, it is a very important element of food both for animals and plants, and in combination with other substances forms a large proportion of animal and vegetable tissues.

CARBON DIOXIDE

Carbonic acid gas, known chemically as CO_2, is a product of combustion. It results from the burning of fuel and is exhaled by the breathing of animals. It also results from certain chemical reactions. The amount in the atmosphere varies slightly, being somewhat greater at night than by day and during cloudy weather than during clear weather. Air containing more than 0.06% of carbon dioxide is not fit to breathe, not because air loaded with carbon dioxide is poisonous, but because it excludes the oxygen and thus produces death by suffocation. It is considerably heavier than air, and in certain localities, where it is emitted from the ground, accumulates in low places in such quantities as to suffocate animals. Death's Gulch, a deep ravine in Yellowstone Park, and Dog's Grotto near Naples, are examples. At the latter place, the gas, on account of being heavier than air, lies so

CO$_2$ is released as a product of combustion

close to the ground that a man, standing erect, will have no difficulty in breathing, while a dog will die of suffocation. It also accumulates in unused wells, cisterns and mines, and can usually be detected by lowering a lighted candle. If carbon dioxide is present in large quantities, the candle will be extinguished because of the lack of oxygen to support combustion.

Although carbon dioxide forms but a small proportion of the atmosphere, it is a very important element in plant life. Animals consume oxygen and exhale carbon dioxide, while plants take in carbon dioxide and give off oxygen; thus, the amount of these gases in the atmosphere is maintained at an equilibrium. Plants, through their leaves, absorb the carbon dioxide, which is decomposed by the sunlight, returning the oxygen free into the air, while the carbon is used to build up plant tissue.

OTHER GASES

Argon, on account of its resemblance to nitrogen, was not discovered until 1894, having been included with the nitrogen in all previous analyses of air. It constitutes about 1% of air by volume. Krepton, neon and xenon exist in minute quantities and have some interest chemically, but little for the meteorologists. Helium and hydrogen probably exist at great elevations in the atmosphere.

Steam rising from hot springs

WATER VAPOR

The vapor of water in the atmosphere varies from about one percent for arid regions to about five percent of the weight of the air for warm, humid regions. It is a little over one-half as heavy as air and moist air is, therefore, lighter than dry air; but the increase of moisture near the center of cyclones has only a slight effect in reducing the pressure. The amount of vapor decreases very rapidly with elevation, and probably disappears at an elevation of five or six miles above the surface. The amount of water in the form of vapor that can exist in the atmosphere increases with the temperature, being .54 grains Troy per cubic foot at zero temperature and 14.81 at 90°. When the air has taken up all the moisture it can contain at a given temperature it is said to be saturated.

The dewpoint is the temperature at which saturation occurs. If the air is saturated, the temperature of the air and the dewpoint will be the same, but if the air is not saturated the dewpoint will be below that of the air.

Relative humidity is expressed in percentages of the amount necessary to saturate. If the air contains one-half enough vapor to saturate it, the relative humidity will be 50%; if one-fourth, enough to saturate, 25%; if saturated 100% etc.

The absolute humidity is the actual amount of water in the form of vapor in the air, and is usually expressed by weight in grains per cubic foot or in inches of mercury, the weight of which would counterbalance the weight of the vapor in the air. The conditions present in a volume of saturated air at a temperature of 32° may be expressed as follows: Relative humidity 100%; dewpoint 32°; absolute humidity 2.11 grains per cu. ft. or .18 inch.

PRESSURE OF ATMOSPHERE

Although the atmosphere is composed of these various gases, it acts in all respects like a simple, single gas. It is very elastic, easily compressed, expands when heated and contracts when cooled. It is acted upon by gravity and, therefore, has weight and exerts pressure, which at sea level amount to about 14.7 pounds on each square inch of the surface. Because it is compressible and has weight, it is more dense at the surface than at any elevation above the surface, and as we ascend in the atmosphere the weight or pressure decreases in proportion to the weight of that part of the atmosphere left below. The weight or pressure of the atmosphere is measured by means of a barometer and is expressed in terms of inches of mercury. The normal atmosphere at sea level will sustain a column of mercury about thirty inches high, and we therefore say that the normal pressure of the atmosphere is thirty inches. (See Lessons on air pressure and the barometer.)

THE HEIGHT OF THE ATMOSPHERE

The air that surrounds the earth is called its atmosphere, but it is a rather curious fact that the earth has really ten atmospheres and may have others not yet discovered.

The air near the surface is a mixture of eight different gases, and each individual gas arranges itself so as to form an atmosphere just as it would if no other gases were present. Thus, the earth is surrounded by an atmosphere of oxygen, an atmosphere of nitrogen, one of carbon dioxide, one of water vapor, one each of argon, krypton, neon, and xenon, while hydrogen and helium are believed to exist at great elevations above the earth's surface.

These gases are kept from flying off into space by the force of grav-

The gases in earths atmosphere scatter blue light more than other wavelengths, giving Earth a blue halo when seen from space

ity, just as a piece of iron, stone, or a building is held fast to the earth by the same force. Gravity acts with greater force on some things than on others. For example, a piece of iron is pulled down by gravity with greater force than is a piece of wood of the same size; likewise, a piece of lead is pulled down with greater force than a piece of iron. We, therefore, say that iron is heavier than wood and that lead is heavier than iron, simply because gravity acts with greater force on the one than on the other. The weight of gases differ just as the weight of different solids, such as lead, wood or iron differ. For instance, nitrogen is 14 and oxygen 16 times heavier than hydrogen.

Gases having the least weight extend upward the farthest, because the lighter the gas the greater its expansive force. Every boy who rides a bicycle takes advantage of the expansive force of air when he pumps his tires. The air is compressed by the pump into the tube and the expansive force exerted by the air in trying to expand makes the tire "stand up." If it requires 10 pounds pressure to compress the gas into the tube, the expansive force will be just ten pounds.

There are two forces in constant operation on each gas that surrounds the earth, viz., expansive force and gravity. Expansive force pushes the gas up and gravity pulls it down, but the force of gravity decreases as the distance from the center of the earth increases, so there is a point at a certain distance above the earth where the two forces just balance each other, and each gas will expand upward to that point but will not rise beyond it. Therefore, if we know the expansive force of a gas and the rate at which gravity decreases, it is possible to calculate the height to which the different gases that compose the air will rise.

In this way it has been determined that carbon dioxide, which is one of the heavier gases, extends upward about ten miles, water vapor about 12 miles, oxygen about 30 miles and nitrogen about 35 miles while hydrogen and helium, the lightest gases known, do not appear at the surface at all, but probably exist at a height of from 30 miles to possibly 200 miles.

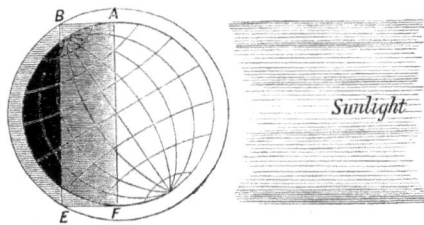

The zone of twilight in midwinter.

There are other ways in which we are able to gain some idea of the approximate height at which there is an appreciable atmosphere. When the rays of light from the sun enter our atmosphere they are broken up or scattered—diffracted—so that the atmosphere is partially lighted for some time before sunrise and after sunset. This is called twilight. If there were no atmosphere, there would be no twilight, and darkness would fall the instant the sun passed below the horizon. Twilight, which is caused by the sun shining on the upper atmosphere, is perceptible until the sun is about 16° below the horizon. From this it is calculated that the atmosphere has sufficient density at a height of 40 miles to scatter, or diffract, sunlight.

Observations of meteors, commonly called shooting stars, indicate that there is an appreciable atmosphere at a height of nearly 200 miles. Meteors are solid bodies flying with great velocity through space. Occasionally they enter our atmosphere. Their velocity is so great that the slight resistance offered by the air generates enough heat by friction, or by the compression of the air in the path of the meteor, to make it red hot or to burn it up before it reaches the bottom of the atmosphere. Only the largest meteors reach the earth.

When a meteor is observed by two or more persons at a known distance from each other, and the angle which the line of vision makes with the horizon is noted by each, it is a simple matter to calculate the distance from the earth where the lines of vision intersect, and thus determine the height of the meteor. In this way, reliable observations have given the height at which there is sufficient density in the atmosphere to render meteors luminous as 188 miles.

TEMPERATURE OF THE ATMOSPHERE

The condition of the atmosphere with respect to its temperature is determined by means of the thermometer. This instrument is in such common use that a detailed description is not necessary. It might be interesting to note that the instrument invented by Galileo was very different from those now in use. Galileo's original thermometer was what is known as an air thermometer, and its operation when subjected to different degrees of heat or cold depended upon the expansion and contraction of air instead of mercury or alcohol. It had one serious defect, viz., the length of a column of air is affected by pressure as well as by temperature and it was, therefore, necessary, when using this thermometer, to obtain the pressure of the atmosphere by means of the barometer before the temperature could be determined. This is obviated in the modern thermometer by the use of mercury or alcohol in a vacuum tube. Mercury is not used when it is expected to register very low temperatures, because it congeals at about 45 degrees below zero Fahr.

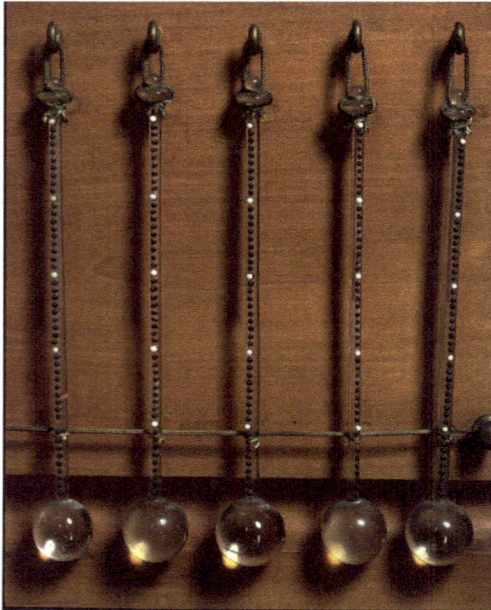

MUSEO GALILEO (CCC BY-SA 4.0)
Early thermometers

THERMOMETER SCALES IN USE

There are three systems in common use for marking the degrees on the scale, viz., Fahrenheit, Centigrade and Reaumur.

The Fahrenheit scale was the invention of a German by that name, but it is worthy of note that this scale is used principally by English speaking nations and is not in common use in Germany. Fahrenheit found that by mixing snow and salt he was able to obtain a very low temperature, and believing that the temperature thus obtained was the lowest possible he started his scale at that point, which he called zero. He then fixed the freezing temperature of water 32 degrees above this zero, and the boiling point of water at 212 degrees. There are, therefore, 180 divisions or degrees between the freezing and boiling point of water on the Fahrenheit scale.

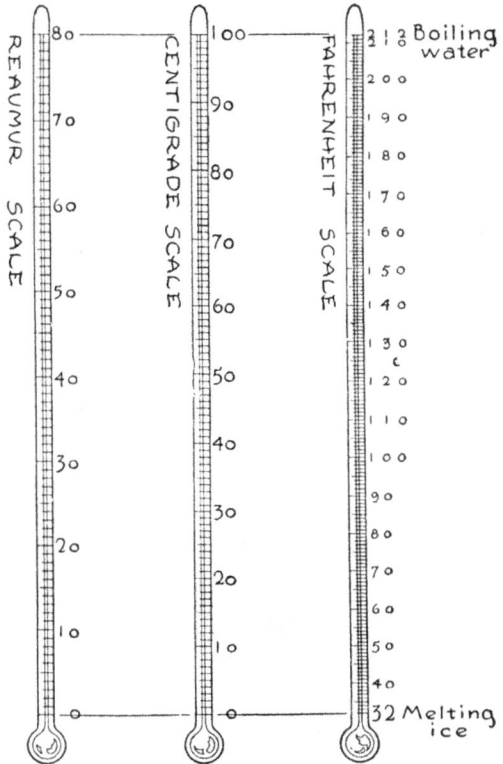

The three standard thermometer scales

The Centigrade scale starts with zero at the freezing point of water and makes the boiling point 100. Thus 180 degrees on the Fahrenheit scale equal 100 degrees on the Centigrade. The Fahrenheit degree is, therefore, only a little more than half as large, to be exact five-ninths of a degree, as a degree on the Centigrade scale. The Centigrade scale is in common use in France and is used almost exclusively in all scientific work throughout the world.

The Reaumur scale is used generally in Russia and quite commonly in some parts of Europe, especially in Germany. On this scale the zero is placed at the freezing point of water and the boiling point at 80 degrees. The divisions are, therefore, larger than those of the Centigrade scale and more than twice as large as the Fahrenheit. The general use of these different scales has led to endless confusion and made the comparison of records difficult, so that even at the present time when making a temperature record it is necessary to indicate the scale in use.

DISTRIBUTION OF THE TEMPERATURE AND PRESSURE

The heat received on the earth from the sun is the controlling factor in all weather conditions. If the earth were composed of all land or all water, and the amount of heat received were everywhere the same throughout the year, there would be no winds, no storms and probably no clouds and no rain, because the force of gravity, which acts on everything on the earth's surface and on the air as well, would soon settle all differences and the atmosphere would become perfectly still. But the earth is composed of land and water and the land heats up more rapidly under sunshine than the water and also gives off—"radiates" its heat more rapidly than water. As a result, the air over the land is warmer in summer than the air over the water. During the winter this is reversed, and the air over the oceans is warmer than the air over the land. The great ocean currents, by carrying the heat from the equatorial regions toward the poles, and by bringing the cold from the polar regions toward the equator, assist in maintaining a constant difference in temperature between the continents and the adjacent oceans.

Furthermore, the fact that the path of the earth about the sun is not a circle but an ellipse, and that the axis of the earth is not perpendicular to the plane of its orbit, result in an unequal distribution of heat over the surface. It is always warmer near the equator than at the poles, and warmer in summer than in winter. All these differences in temperature cause corresponding differences in density, which, in turn, cause differences in weight or pressure over various parts of the earth's surface. These changes are, in no way, the result of chance but are determined by the operation of fixed natural laws, and with this in mind we may now take up the study of the winds of the world.

THE WINDS OF
THE WORLD

The general circulation of the atmosphere may be best studied by disregarding those smaller differences of temperature and pressure that result from local causes and by viewing the earth and its at-

Strong winds

mosphere as a whole, considering only those larger differences which are in constant operation. In the great oceans of the world we find the water constantly moving in a very systematic manner, and we call this system of movements ocean currents. The Gulf Stream, the Equatorial Current, the Japan Current and others may be likened to great rivers of water moving systematically on their courses in the ocean.

There are greater rivers of air in the atmosphere than any in the oceans, and they move on their courses with equally systematic precision and in obedience to fixed laws, which we may in a measure understand.

The river, at the bottom of which we live, is broad and deep, extending in width from Florida northward nearly to the north pole. It flows from west to east circling the globe and its name is The Prevailing Westerlies. The other river in this hemisphere extends southward from latitude about 35° nearly to the equator. Its name is The Northeast Trade Winds.

In the southern hemisphere are two similar rivers, one extending southward from latitude about 30° nearly to the south pole with its current, like its counterpart in the northern hemisphere, flowing from west to east, circling the globe. It is also called The Prevailing Westerlies. The other river in the southern hemisphere extends from about latitude 30° northward nearly to the equator and flows from the southeast toward the northwest, hence the name Southeast Trade Winds. The dividing line, or bank, between the rivers in each hemisphere belts the earth at about 35° north and 30° south of the equator.

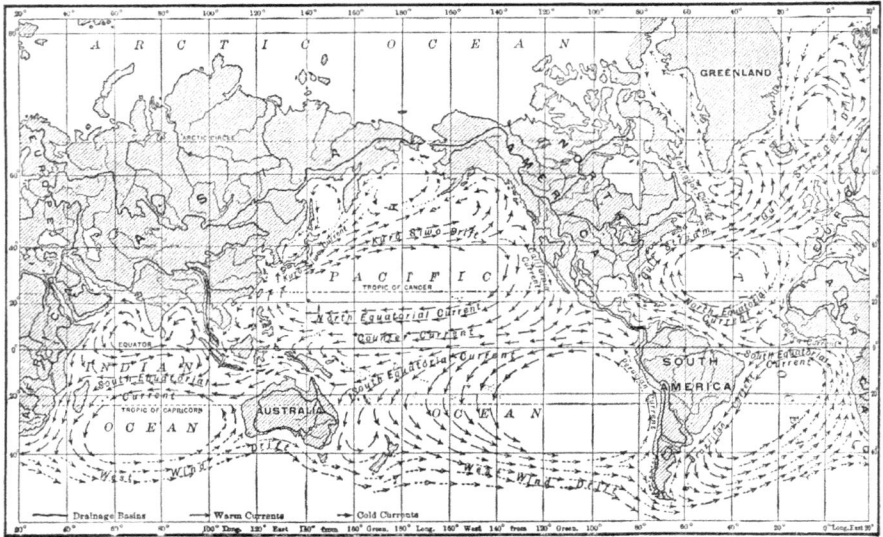

Why does the air move and why does it move in such a regular, systematic manner? To answer these questions we will rely upon gravity, the heat from the sun and the effect of the rotation of the earth on moving wind currents.

Everyone knows that water flows down hill because of the force of gravity. Gravity is nature's great peacemaker. It is always trying to settle disturbances, even things up, smooth them over. If there were no winds to bring rain to the land or to stir up the ocean, gravity would soon run all the water into the lakes and the seas, and then smooth them out like sheets of glass; and if there were nothing to stir up the winds, gravity would soon settle all differences in the atmosphere and the air would become perfectly quiet. So gravity is kept busy trying to smooth out the water which the wind stirs up, at the same time trying to quiet the winds which are stirred up by the heat of the sun.

Tyndall says that heat is a mode of motion, that when heat is imparted to a substance the molecules of which it is composed are set into very rapid vibration. They are continually trying to get away from each other and usually succeed in getting more space, and thus increase the size or volume of the substance, or, in other words, expand it. Iron, brass, copper, water and many other substances expand under heat. Air is a gas and expands very rapidly when heated. One cubic foot of cold air becomes two cubic feet when heated. Now gravity pulls things down toward the center of the earth in accordance with

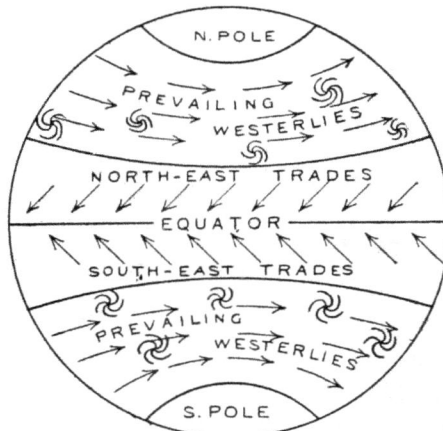

WINDS OF THE WORLD

their weight-density, and a cubic foot of cold air, being more dense and thus heavier than an equal volume of warm air, is pulled down with greater force. We, therefore, say that warm air is lighter than cold air, and if lighter it will rise. What it actually does is to press equally in all directions and when a place is found where there is less resistance than elsewhere it moves in that direction. So when heat causes air to expand and become lighter than the surrounding cool air, it moves, and air in motion is *wind*.

This diagram represents a section of the atmosphere over a broad, level plain with the air at rest and pressing down equally on every part of the surface. The dotted line H represents the top of the quiet atmosphere. Such a condition occurs frequently at night after the heat from the sun is withdrawn and gravity has settled the atmosphere. When the rays of the sun fall upon the earth upon which this quiet air rests they warm the earth first, then the layer of air immediately in contact with the surface, so the atmosphere is heated from the bottom upward. We will assume that the layer of air between the earth and the dotted line, G, is thus heated to a higher temperature than the air above it. It will, therefore, expand. It cannot expand downward because of the earth. It cannot expand much laterally because it is pressed upon by air that is also seeking more space. It, therefore, expands upward as represented by the line A B C. Now in expanding upward it lifts all the air above it and the line H, representing the top of the atmosphere, will become bowed upward also as indicated by the

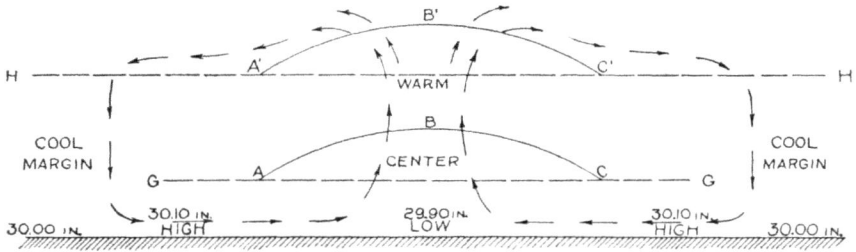

Fig. 1. Diagram showing air currents set up by sun's heat

line A'B'C'. As a result, the air at the top of the atmosphere over the warm center slides down the slopes on either side toward the cool margins. As soon as the flow of air away from the warm center begins, just that instant the pressure upon the heated layer at the surface is relieved and the warm air rushes upward (is pushed upward) and the whole circulation, as indicated by the arrows, begins. It must be remembered that gravity is the really active force in maintaining this movement, because it pulls down the denser, heavier air at the cool margins with greater force than the warm, expanded, light air at the warm center. The descent of the cool air actually lifts the warm air.

The normal pressure, or weight, of the atmosphere at sea level is about 14.7 pounds on each square inch of surface. It is customary, however, to express the weight of the atmosphere in terms of inches of mercury instead of in pounds and ounces. A column of air one inch square from sea level to the top of the atmosphere will just counterbalance a column of mercury 30.00 inches high in a barometer tube of the same size. We, therefore, say that the normal pressure of the atmosphere at sea level is about 30.00 inches. If, for any reason, the atmosphere becomes heavier than normal, it will raise the column of mercury above the 30 inch mark, and we say that the pressure is "high." If the atmosphere becomes lighter than normal, we say that the pressure is "low." So high pressure means a heavy atmosphere and low pressure a light atmosphere.

At the beginning we assumed that the atmosphere over the broad, level plain was quiet and that it pressed down equally on every part of the surface. We will now assume that the pressure was normal, or 30.00 inches, and note the changes in pressure that result from the interchange of air between the warm center and the cool margins. So long as none of the air raised by the expanding layer at the surface, moved away toward the cool margins, no change in pressure occurred; but the instant the air began to glide down the slopes away from the warm center, then the pressure at the surface decreased, because, some air having moved away, there was less to press down than before. The pressure at the warm center, therefore, became less than 30.00 inch-

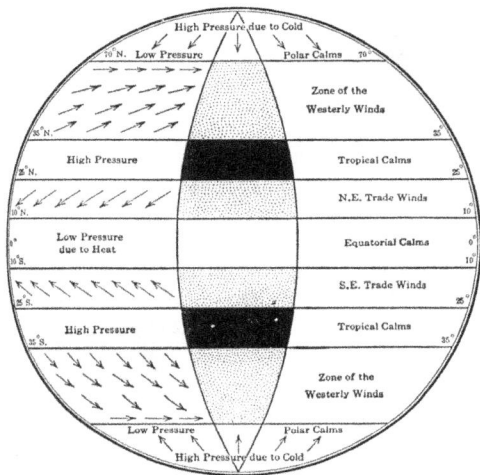

Pressure belts on a simplified globe

es, or in other words, low. Likewise, the air as it moved away from the warm center, having lost much of its heat during its ascent, was gradually pulled down by gravity because of its greater density, thus increasing the pressure over the cool margins. We, therefore, have low pressure at the warm center, 29.90 inches and high pressure, 30.10 inches, at the cool margins. From this illustration we obtain the six principles of convectional circulation, viz.:

1. Low pressure at warm center.
2. High pressure at cool margins.
3. Ascending currents at warm center.
4. Descending currents at cool margins.
5. Surface winds from high pressure to low pressure.
6. Upper currents from low pressure to high pressure.

Now, we all know that the temperature of air is much higher at the equator than at the poles and we may, therefore, let Figure 1 represent a section of the atmosphere along any meridian from the north to the south pole. The equator would then become the warm center and the poles the cool margins. We would then expect to find a belt of low pressure around the world near the equator because of the high temperature, and high pressure at the poles because of the low tempera-

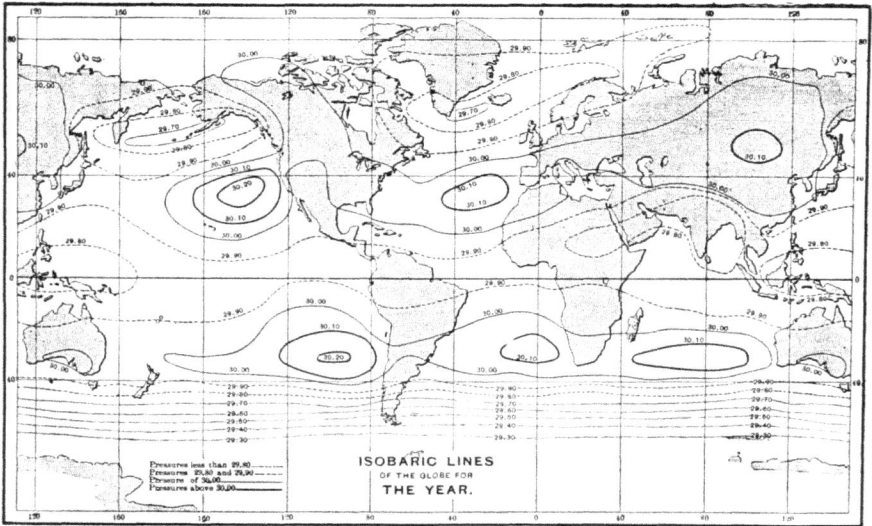

Figure 2. Isobars of the world.

ture. We would, also, expect to find ascending currents at the equator; upper currents flowing from the equator toward the poles; descending currents at the poles, and surface winds blowing from the poles toward the equator. Let us now test our theory by actual facts and see how far they are in accord.

The chart, Figure 2, represents the normal, or average, pressure at sea level for the world, and if our theory is in accord with the facts, we should find a belt of low pressure all around the world near the equator, with areas of high pressure at the poles. Let us examine the chart. Beginning at the equator, and bearing in mind that the normal pressure is about 30.00 inches, we find irregular lines, representing pressures of 29.90 inches—slightly below normal—around the world on both sides of the equator. Between these lines we find pressure as low as 29.80. It is, therefore, evident that there is a belt of low pressure around the world near the equator, as anticipated. Let us look for the high pressure at the poles. We have comparatively few observations near the poles, but the line nearest the south pole is marked 29.30 inches, a surprisingly low pressure, much lower even than the low belt at the equator, and just the reverse of what we expected to find. When we look at the north pole we find that the pressure is not so low as at the south pole, but still below normal and about as low as

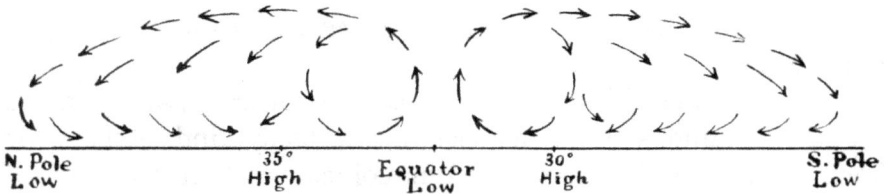

Figure 3. Diagram showing air currents along any meridian.

at the equator. Going north and south from the equator we find that the pressure increases gradually up to about latitude 35° in the northern hemisphere and to about latitude 30° in the southern, after which it decreases toward the poles. So there are two well marked belts of high pressure circling the globe; the one about 35° north, and the other about 30°, south of the equator. May it not be significant that these belts of high pressure coincide so nearly with the margins, or banks, of the air rivers mentioned earlier?

Thus far our theory does not accord very well with the facts. True, we found the low pressure at the equator as anticipated; but we also found low pressure at the poles, where the reverse was expected; and the high pressure that we anticipated at the poles, we found not far north and south of the equator. We will, therefore, have to discard our theory, or reconstruct it to accord with the facts. Let us reconstruct Figure 1, and mark the pressure on the line representing the earth's surface along any meridian to accord with the facts as they appear on Figure 2.

The above diagram now represents the true pressure along any meridian, as determined by actual observations, and we cannot escape the conviction that the requirements as to temperature and pressure at the warm center are fulfilled by the high temperature and low pressure found at the equator. Furthermore, the temperature decreases north and south from the equator, and thus the belts of high pressure near the tropics may be taken to represent the conditions at the cool margins. The first and second principles of a convectional circulation, viz., low pressure at the warm center and a high pressure at the cool margins, are thus fulfilled. To satisfy the remaining conditions, we should find ascending currents near the equator, upper currents flowing from the equator toward the tropical belts of high pressure, descending currents at the tropics, and surface winds blowing from the tropics toward the equator. Let us now examine the surface winds of the world.

On either side of the equator and blowing toward it, we find the famous trade winds—the most constant and steady winds of the world. Their northern and southern margins coincide with the tropical belts of high pressure. They blow from high pressure to low pressure and we cannot doubt that they act in obedience to the fifth principle of convectional circulation. From observation of the lofty, cirrus clouds in the trade wind belts, we have abundant evidence of upper currents, flowing away from the equator toward the tropical belts of high pressure; thus the sixth principle is satisfied. The torrential rains and violent thunderstorms, characteristic of the equatorial regions, bear evidence to the rapid cooling of the ascending currents near the equator; while the clear, cool weather and light winds of the Horse Latitudes clearly indicate the presence of descending currents at the tropics. Thus, the six principles of a convectional circulation are satisfied, and the evidence is conclusive that the trade winds form a part of a convectional circulation between the tropical belts of high pressure and the equatorial belt of low pressure.

Cup anemometer. The dial cover is removed to show the mechanism

You have doubtless observed that the trade winds do not blow directly toward the equator but are turned to the west so that they blow from the northeast in the northern hemisphere, and from the southeast in the southern. This peculiarity is not in strict accord with our ideas of a simple convectional circulation and suggests, at least, the presence of some outside influence. If we turn to Ferre's treatise on the winds, we find a demonstration of the following principle: a free moving body, such as air, in moving over the surface of a rotating globe, such as the earth, describes a path on the surface that turns to the right of the direction of motion in the northern hemisphere and to the left in the southern. The curvature of the path increases with the latitude, being zero at the equator and greatest at the poles, and is independent of direction. With this in mind, if we take position at the northern limit of the trade winds in the northern hemisphere and face the equator, we find that the winds moving toward the equator turn to our right; likewise, if we face the equator from the southern limit of the southeast trades, we find them turning to our left. Observations of upper clouds in the trade wind belts show that the upper currents also turn to the right in the northern hemisphere, and to the left in the southern. It is, therefore, clear that the systematic turning of the trade winds from the meridian is due to the rotation of the earth. The value of a force at various latitudes and for various velocities that would cause a body to turn away from a straight line, is purely a problem in mathematics, and for the benefit of those versed in the science the formula is given. The amount of such a force is expressed by 2 MVW sin D, where M is the mass, V the velocity, W the angular rotation of the earth, and D the latitude.

Not all of us may be able to solve the problem, but we may understand something of the effect of the rotation of the earth on moving wind currents. It is a well-known principle of physics that if a body be given a motion in any direction, it will continue to move in a straight line by reason of its inertia, without reference to north, south, east or west. A personal experience of this principle may be gained in a street car while it is rounding a curve.

In this diagram, we have a view of the northern hemisphere. The direction of the rotation is indicated by the curved arrows outside the circle representing the equator. Suppose that a wind starts from the equator, moving along the meridian A directly toward the north pole. It is clear that it cannot continue to move along the meridian, because the direction of the meridian with reference to space, is continually changing, and the inertia of the wind compels it to move in a straight line without reference to the points of the compass. So when the meridian A has been moved to B by the rotation of the earth, the wind, although it maintains its original direction, no longer points toward the pole but to the right of the pole. Likewise, a wind starting from the pole toward the equator also turns to the right of the meridians and becomes a northeast wind as it approaches the equator. A wind moving east or west, also turns to the right of the parallels for the same reason. So a wind starting out from the equator with the best possible intention of hitting the pole, and all the while continuing in the same straight line, will miss the pole by many miles, and always on the right side in the northern and on the left side in the southern hemisphere. Thus, the oblique movement of both the trade winds and the prevailing westerlies is accounted for.

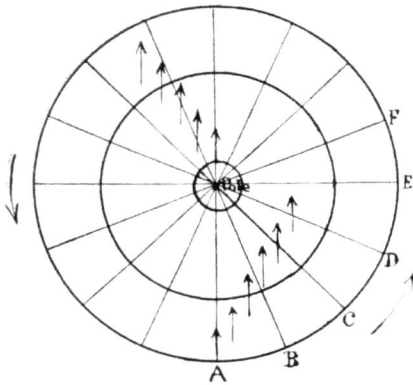

Diagram showing effect of earth's rotation on the atmosphere.

It now remains to consider the cause of the unexpected low pressure found at the poles, and the reason for the belts of high pressure at the tropics. If we refer to Figure 2, it is evident that not all the air that ascends at the equator descends at the tropics, else there would be an absence of air at the higher latitudes, which is manifestly not the case. On the other hand, it is equally impossible that all the air ascending at the equator should move to the poles, because the space it could occupy decreases rapidly from a maximum at the equator to zero at the poles. Only a part of the air that ascends at the equator is, therefore, involved in the trade wind circulation and a part passes over the tropics and moves on toward the low pressure at the poles. Furthermore, some of the air that descends at the tropics moves along the surface toward the poles, obeying the law that impels air to move from high pressure to low pressure. Now every particle of air that passes over the tropics, every particle that moves northward along the surface, turns to the right in the northern and to the left in the southern hemisphere. All, therefore, miss the poles—on the right side in the northern and on the left side in the southern hemisphere. The result is that two great whirlpools develop in the atmosphere; one whirling about the north

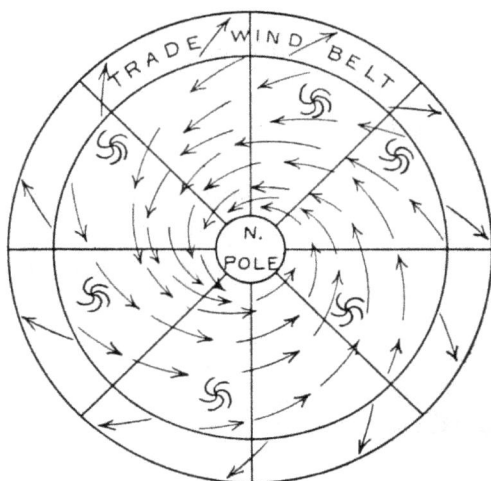

CIRCUMPOLAR WHIRL

and the other whirling about the south pole. The outer margins of these whirlpools coincide with the tropical belts of high pressure.

As an example of a whirlpool we may take a basin having a vent at the center of the bottom. If the basin is filled with water, the plug withdrawn and the water given a slight rotary motion, its velocity will increase as it approaches the center and the rapid whirling will develop sufficient centrifugal force to open an empty core. Those who have visited the great whirlpool at Niagara, undoubtedly noticed that the whirling waters are held away from the center and piled up around the margins by the centrifugal force developed. Let us suppose that air starting from the equator, moves without friction or other resistances toward the pole. Its velocity must increase as its radius shortens, because the law of the conservation of areas requires that the radius must always sweep over equal areas in a given unit of time. (See law of conservation of areas.) At the equator, the air has an easterly motion equal to the eastward motion of the earth, which is 1,000 miles per hour. At latitude 60° the radius will have decreased one-half and the velocity, therefore, doubled; but at latitude 60° the eastward motion of the earth is only 500 miles per hour, so the air would be moving 1500 miles per hour faster than the earth. At a distance of 40 miles from the pole the wind would attain an easterly velocity of 100,000 miles per hour, and moving on so short a radius would develop sufficient centrifugal force to hold all the air away from the pole and thus form a vacuum. That the supposed case of no friction is far from the truth is evidenced by the fact that the pressure at the north pole is but little less than at the equator; but the centrifugal force developed by the gyration winds, in thus withdrawing the air from the poles and piling it up at the tropics, may be fairly taken as sufficient cause for the low pressure found at the poles and the belts of high pressure at the tropics.

The questions that remain to be considered are: (1) the low pressure at the south pole as compared with the pressure at the north pole and (2) the unequal distance of the tropical belts of high pressure from the equator. These questions may be considered together.

It is to be remembered that the southern hemisphere is the water hemisphere, and that the prevailing westerlies, in gliding over the smooth water surface, are but little retarded by friction and, therefore, attain a higher velocity than the corresponding winds of the northern hemisphere, where the rougher surface materially retards their movement. As a consequence, the circumpolar whirl of the southern hemisphere is stronger, and develops a greater centrifugal force, thus holding a larger quantity of air away from the south pole and reducing the pressure to a greater degree than is brought about by the weaker winds of the northern hemisphere.

Since the circumpolar whirl of the southern hemisphere is the stronger of the two, it withdraws the air to a greater distance from the pole than does its weaker counterpart of the northern hemisphere, and piles it up in the tropical belt of high pressure about five degrees nearer the equator than does the weaker forces of the northern hemisphere.

STORMS

Having gained a comprehensive view of the general, planetary wind system, we may now undertake the study of local disturbances that arise within the general circulation and are known as "storms."

Storms are simply eddies in the atmosphere. They may be com-

A huge storm system

pared to the eddies that are often seen floating along with the current of a river or creek. In these eddies the water is seen to move rapidly around a central vertex, developing sufficient centrifugal force to hold some of the water away from the center, thus forming a well marked depression, frequently of considerable depth. The whole circulation of the eddy is quite independent of the current of the stream which carries it along its course, and while its general direction and velocity of movement coincide with that of the current, there are times when it will be seen to move quickly from side to side and again when it will remain nearly stationary for a time or take on a rapid movement.

The eddies or storms in the atmosphere act in much the same way. They are carried along by the general currents of the river of air in which they exist. Their general direction coincides with the direction of the current in which they are floating, and their rate of movement conforms in a general way to its velocity; but like the eddies in the river, they do not always move in straight lines nor at a uniform rate of speed.

There is one important respect in which the eddies in the air differ from eddies in water. The water eddy may revolve in either direction, depending upon the direction in which the initial force was applied, but the storm eddies in the atmosphere always revolve counter-clockwise in the northern hemisphere, and clockwise in the southern.

This is due to the deflecting force of the earth's rotation.

WEATHER MAPS

A weather map is a sort of flashlight photograph of a section of the bottom of one or more of these great rivers of air. It brings into view the whole meteorological situation over a large territory at a given instant of time; and, while a single map conveys no indication of the movements continually taking place in the atmosphere, a series of maps, like a moving picture, shows not only the whirling eddies, the hurrying clouds and the fast-moving winds, but the ceaseless on-flow of the great river of air in which they float. Our present knowledge of the movements of the atmosphere has been gained chiefly from a study of weather maps; they form the basis of the modern system of weather forecasting, and their careful study is essential to any adequate understanding of the problems presented by the atmosphere. (See page 105.)

THE PRINCIPLES OF WEATHER FORECASTING

The forecasting of the weather has been made possible by the electric telegraph. It is based upon a perfectly simple, rational process constantly employed in everyday affairs. We go to a railway station and ask the operator about a certain train. He tells us that it will arrive in an hour. We accept his statement without question, because we are confident that he knows the speed at which the train is approaching, a few clicks of his telegraph instrument has told him just where it is and the time it will arrive, barring accidents, is a simple calculation. Information of coming weather changes are obtained in a similar manner. Although storms do not run on steel rails like a train, nevertheless their movements may be foreseen with a reasonable degree of accuracy, depending chiefly upon the size of the territory from which telegraphic reports are received and the experience and skill of the forecaster. As a rule, the larger the territory brought under observation, especially in its longitudinal extent (the general currents carry storms of the middle latitudes eastward around the world and those of the tropics westward), the earlier advancing changes may be recognized and the more accurately their movements foreseen.

FORECASTS BASED ON WEATHER MAPS

The forecasts issued by the United States Weather Bureau are based on weather maps, prepared from observations taken at 8 a.m. and 8 p.m. at about 200 observatories. In addition to the reports received by telegraph by the Central Office at Washington, the several forecast centers and other designated stations from observatories or stations in the United States, a system of interchange with Canada, Mexico, the West Indies and other island outposts in the Atlantic and Pacific gives to the forecaster two daily photographs of the weather conditions over a territory embracing nearly the whole of the inhabited part of the western hemisphere north of the equator. Any sort of disturbance within this vast region is photographed at once upon the weather map. If it be a West Indies hurricane or other destructive storm, its character is recognized instantly, its rate and direction determined and information of the probable time of its arrival sent to those places that lie in its path. The method is perfectly simple. Anyone with a weather map and a little experience can forecast the weather with some degree of accuracy, or, at least, gain an intelligent understanding of the conditions upon which the forecasts that accompany the map are based.

Maps, Where Published and How Obtained

Weather maps are published in many daily papers, and in somewhat larger form and more in detail, at many Weather Bureau stations. They may usually be obtained for school use by applying to the nearest Weather Bureau station or to the Chief of the Weather Bureau at Washington, D. C.

The forecasts that accompany the maps are simply an expression on the part of the official forecaster as to the weather changes he expects to occur in various parts of the country within the time specified, usually within 36 to 48 hours. His opinion is based upon the conditions shown by the map. He has no secret source of information. You may accept his conclusions, or, if in your opinion they are not justified, you have all the information necessary to make a forecast for yourself. Weather maps are published so extensively with a view to thus stimulating an intelligent interest in the problem of weather forecasting, and also that one may see at a glance what the temperature, rainfall, wind and weather is in any part of the country in which he may be interested. The friends of the weather service are those who best understand its work.

The Value of the Weather Service

No one knows so well as the forecaster that the changes that appear most certain to come sometimes fail, or come too late; but taking all in all, about 85 out of 100 forecasts are correct. Of those that fail, probably not more than three or four per cent. fail because the changes come unannounced. Most forecasters predict too much, and their forecasts fail because the expected changes come after the time specified or not at all. It is fortunate that this is so; for it is better to be prepared for the change though it be late in coming than to have it come without warning.

The value of the weather service to the agriculture and commerce of the United States cannot be questioned seriously. That the appropriations for its support have been increased year by year from $1,500 in 1871 to nearly $1,500,000 in 1910 is evidence of its value and efficiency. A conservative estimate places the value of property saved by the warnings issued by the Weather Bureau at $30,000,000 annually.

Experiments To Show Air Pressure

Leading thought— The air presses equally in all directions.

Experiment 1—To show that air presses upward— Fill a tumbler which has an unbroken edge as full of water as possible. Take a piece of writing paper and cover the tumbler, pressing the paper down firmly upon the edge of the glass. Turn the glass bottom side up and ask why the water does not flow out. Allow a little air to enter; what happens? Why? Turn the glass filled with water and covered with paper sidewise; does the water flow out? If not, why?

Experiment 2—To show that air passes downward— Ask some of the boys of the class to make what they call a sucker. This is a piece of leather a few inches across. Through its center a string is drawn which fits very closely into the leather and is held in place by a very flat knot on the lower side. Dampen the leather and press it against any flat surface, and try to pull it off. If possible, place the sucker on a flat stone and see how heavy a stone can be lifted by the sucker. Ask why a sucker clings so to the flat surface. If a little air is allowed to get between the sucker and the stone, what happens? Why?

JEAN-JACQUES MILAN (CC BY-SA 3.0)
A traditional barograph, without its protective case.

Hints to the teacher regarding the Experiments— The water is kept in the tumbler in Experiment 1 by the pressure of the atmosphere against the paper. If the tumbler is tipped to one side the water still remains in the glass, which shows that the air is pressing against the paper from the side with sufficient force to restrain the water, and if the tumbler is tipped bottom side up it shows the air is pressing upward with sufficient force to keep the water within the glass.

In the case of Experiment 2, we know that the leather pressing upon the floor or on the stone is not in itself adhesive, but it is made wet simply so that it shall press against the smooth surface more closely. The reason why we cannot pull it off is that the air is pressing down upon it with the force of about fifteen pounds to the square inch. If the experiment is performed at sea level, we should be able to lift by the string of the sucker a stone weighing fifteen pounds. The reason why the water falls out of the tumbler after a little air is let beneath the paper, is that then the air is pressing on both sides of the paper; and the reason why the sucker will not hold if there is any air between it and the stone, is because the air is pressing in both directions upon it.

Supplementary reading— The Wonderbook of the Atmosphere, Houston, Chapters III, IV, V.

Plastic bottle sealed at 14,000 feet (4267 m) on Mauna Kea observatory on the island of Hawaii, taken down to 9000 feet (2743 m) and then 1000 feet (305 m), where the change in air pressure had crushed it.

The Barometer

LESSON

Leading thought— The weight of our atmosphere balances a column of mercury about thirty inches high, and is equal to about fifteen pounds to the square inch. This pressure varies from day to day, and becomes less as the height of the place increases. The barometer is an instrument for measuring the atmospheric pressure. It is used in finding the height of mountains, and, to a certain extent, it indicates changes of the weather.

Method— A glass tube about 36 inches long, closed at one end; a little glass funnel about an inch in diameter at the top; a small cup—a bird's bathtub is a good size since it allows plenty of room for the fingers; mercury enough to fill the tube and have the mercury an inch or more deep in the cup. Be careful not to spill the mercury in the following process, or you will be as badly off as old Sisyphus with his rolling stone.

Set the closed end of the tube in the cup so that any spilled mercury will not be lost; with the help of the funnel slowly and carefully fill the tube clear to the top with the mercury; empty the rest of the mercury into the cup; place the end of one of the fingers of the left hand tightly over the open end of the tube and keep it there; with the right hand invert the tube, keeping the end closed with the finger, and place the hand, finger and all, beneath the mercury in the cup, then remove the finger, keeping the open end of the tube all the time below the surface of the mercury. When the mercury has ceased to fall measure the distance from the surface in the cup to the top of the mercury in the tube.

Observations—

1. How high is the column of mercury in the tube?

A barometer made by pupils

2. What keeps the mercury in the tube? Place the cup and the tube on a table in the corner of the room, place behind the tube a yardstick, and note whether the column of mercury is the same height day after day. If it varies, why?

3. Would the mercury column be as high in the tube if it were placed on top of a mountain as it would at the foot? Why?

Supplementary reading— Chap. II in *The Wonderbook of the Atmosphere,* Houston.

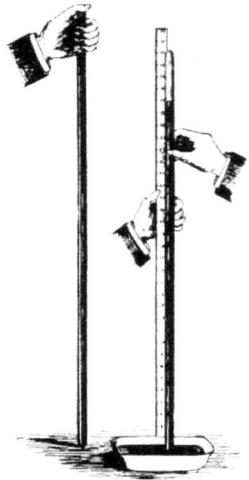

How To Read Weather Maps

WEATHER maps may be obtained by writing to the nearest Weather Station, or by writing to the Chief of the Weather Bureau, Dr. Willis L. Moore, Washington, D. C., stating that you wish to post the maps in a public place. A supply of maps for three successive days for use in these lessons may be obtained at 20 cents per hundred. Sometimes they are sent free, if it is stated that they are to be used for school purposes.

The words isobar and isotherm have been bogies which have frightened many a teacher from undertaking to teach about weather maps, and yet how simple are the meaning of these two words. Isobar is made up of two Greek words, *isos* meaning equal and *baros* meaning weight. Therefore, an isobar means equal weight, and on a map one of these continuous lines means that, wherever it passes, the atmosphere there has equal weight and the barometer stands at equal height. The isobar of 30 means that the mercury in the barometer stands 30 inches in height in all the regions where that line passes.

"Isotherm" comes from the two Greek words, *isos* meaning equal and *therme* meaning heat. Therefore, on the map the dotted lines show the region where the temperature is the same. If at the end of the dotted line you find 60 it means that, wherever that line passes, the thermometer stands at 60 degrees.

Many of the "highs" and "lows" enter the United States from the Pacific Ocean about the latitude of Washington State or southwest British Columbia. They follow one another alternately, crossing the continent in the general direction of west to east in a path which curves somewhat to the north, and they leave the United States in the latitude of Maine or New Brunswick. If they enter by way of lower California, they pass over to the Atlantic Ocean farther south. The time for the passage of a high or low across the continent averages about three and one-half days, sometimes a little more. These areas are usually more marked in winter, and wind storms are more marked and more regular.

A low area is called a cyclone and a high area an anti-cyclone. The destructive winds, popularly called cyclones, which occur in certain regions, should be called tornadoes instead, although in fact they are simply small and violent cyclones. But a cyclone, when used in a meteorological sense, extends over thousands of square miles and is not violent; while a tornado may be only a few rods in diameter and be very destructive. The little whirlwinds which lift the dust in the roads are rotary winds also, but merely the eddies of a gentle wind.

In a cyclone or "low," and also in a tornado, the air blows from all sides spirally inward *toward* the center where there is a column of *ascending* air.

In an anti-cyclone or "high" the air blows outward in every direction in curved lines *from* a column of *descending* air.

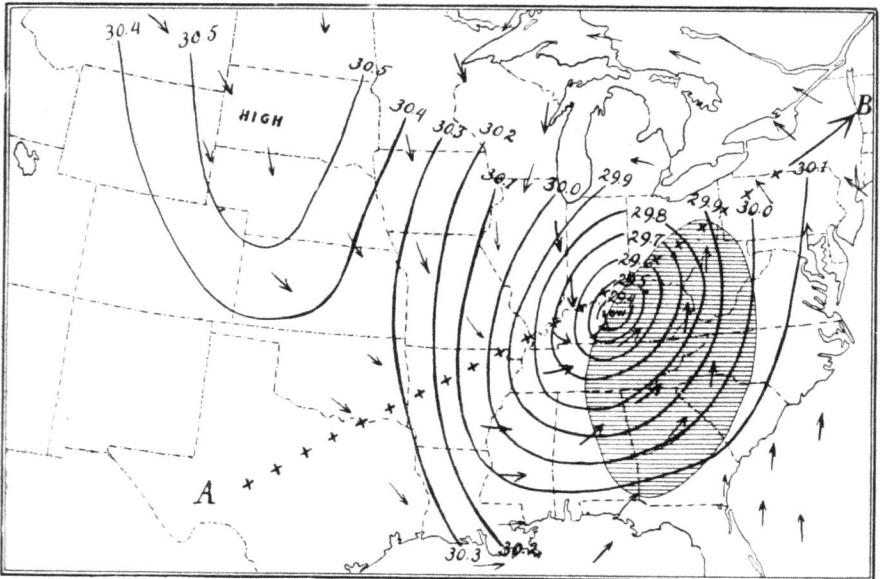

Map of a storm

In the above map, the curved lines are isobars; the line of crosses from A to B indicates the course of the storm; the arrows indicate the direction of the wind, note that it is moving counter-clockwise around the area of low pressure; the shaded area indicates the region where it is raining or snowing; note that this is the area where the warm, moist

DEPARTMENT OF AGRICULTURE, WEATHER BUREAU

EXPLANATION OF WEATHER SIGNALS

No. 1	No. 2	No. 3	No. 4	No. 5
Fair Weather	Rain or Snow	Local Rain or Snow	Temperature	Cold Wave

INTERPRETATION OF DISPLAYS

No. 1, alone, indicates fair weather, stationary temperature.
No. 2, alone, indicates rain or snow, stationary temperature.
No. 3, alone. indicates local rain or snow, stationary temperature.
No. 1, with No. 4 above it, indicates fair weather, warmer.
No. 1, with No. 4 below it, indicates fair weather, colder.
No. 2, with No. 4 above it, indicates rain or snow, warmer.
No. 2, with No. 4 below it, indicates rain or snow, colder.
No. 3, with No. 4 above it, indicates local rain or snow, warmer.
No. 3, with No. 4 below it, indicates local rain or snow colder.

WILLIS L. MOORE
Chief U.S. Weather Bureau.

EXPLANATION OF WHISTLE SIGNALS

A warning blast of fifteen to twenty seconds duration is sounded to attract attention. After this warning the longer blasts (of four to six seconds duration) refer to weather, and shorter blasts (of one to three seconds duration) refer to temperature; those for weather are sounded first.

Blasts	Indicate	Blasts	Indicate
One long	Fair weather.	One short	Lower temperature.
Two long	Rain or snow.	Two short	Higher temperature.
Three long	Local rain or snow.	Three short	Cold wave.

By repeating each combination a few times, with intervals of ten seconds, liability to error in reading the signals may be avoided.

STORM AND HURRICANE WARNINGS

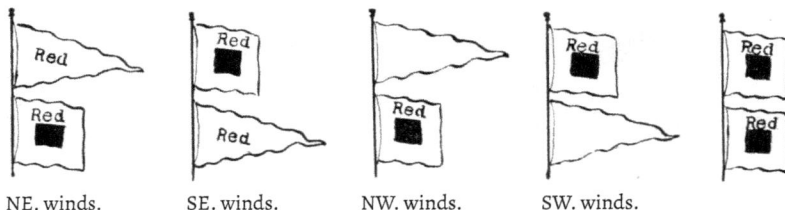

Storm warnings.   Hurricane warning.

NE. winds.	SE. winds.	NW. winds.	SW. winds.	

EXPLANATION OF STORM AND HURRICANE SIGNALS

Storm warning--A red flag with a black center indicates that a storm of marked violence is expected. The pennants displayed with the flags indicate the direction of the wind; red, easterly (from northeast to south); white (westerly from southwest to north). The pennant above the flag indicates that he wind is expected to blow from the northerly quadrants: below from the southerly quadrants.

By night a red light indicates easterly winds and a white light below a red light westerly winds.

Hurricane warning--Two red flags with black centers displayed one above the other indicates the expected approach of a tropical hurricane or one of those extremely severe and dangerous storms which occasionally move across the Lakes and northern Atlantic coast.

No night hurricane warnings are displayed.

SATURDAY, DECEMBER 24, 1904 - 8 A.M.

SUNDAY, DECEMBER 25, 1904, 8 A.M.

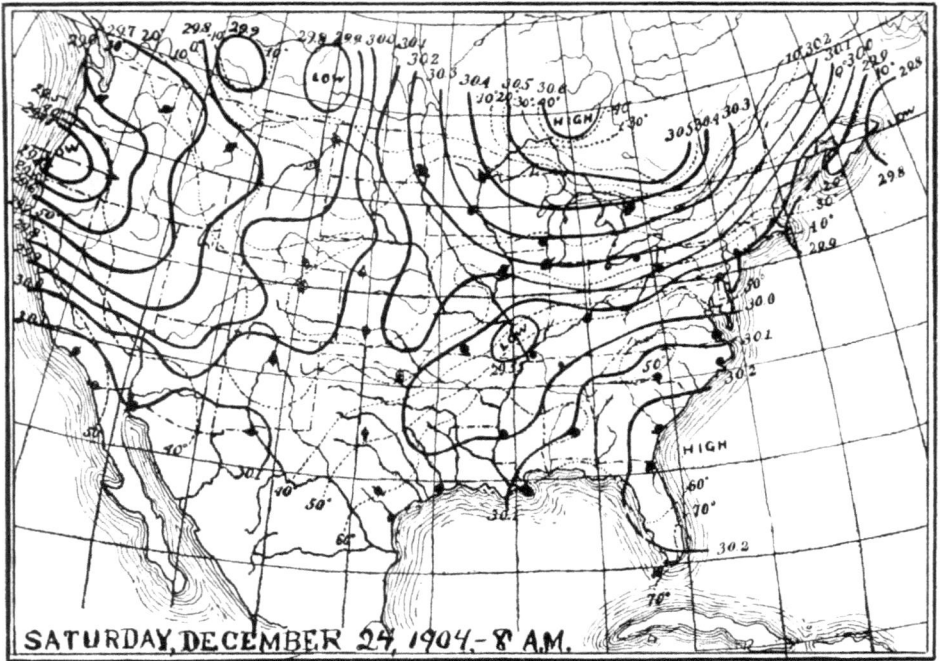

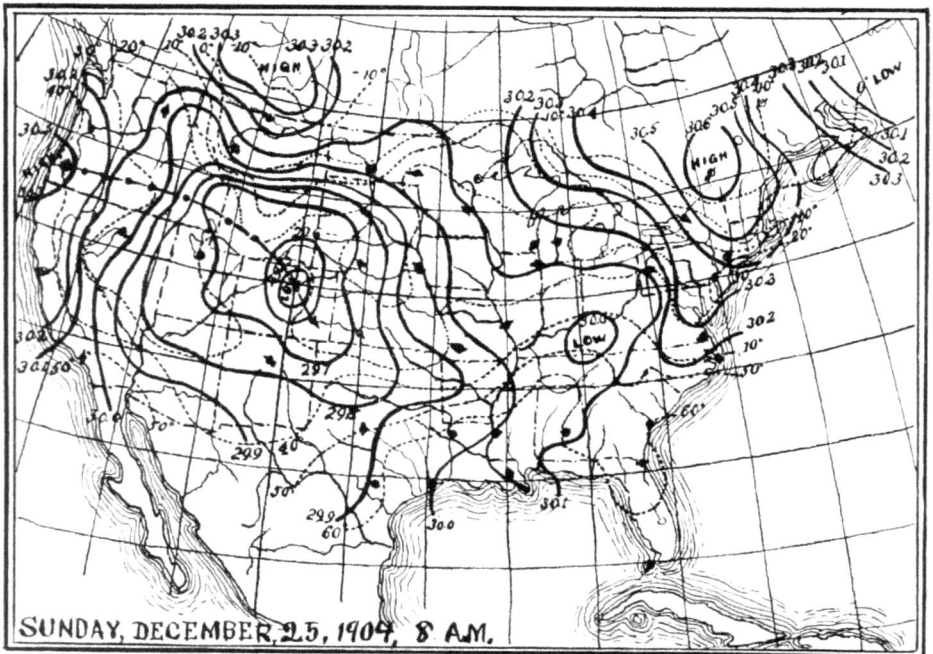

U.S. weather maps, showing the eastward progress
Note the course of the low that was on the Pacific Coast Dec. 4; this

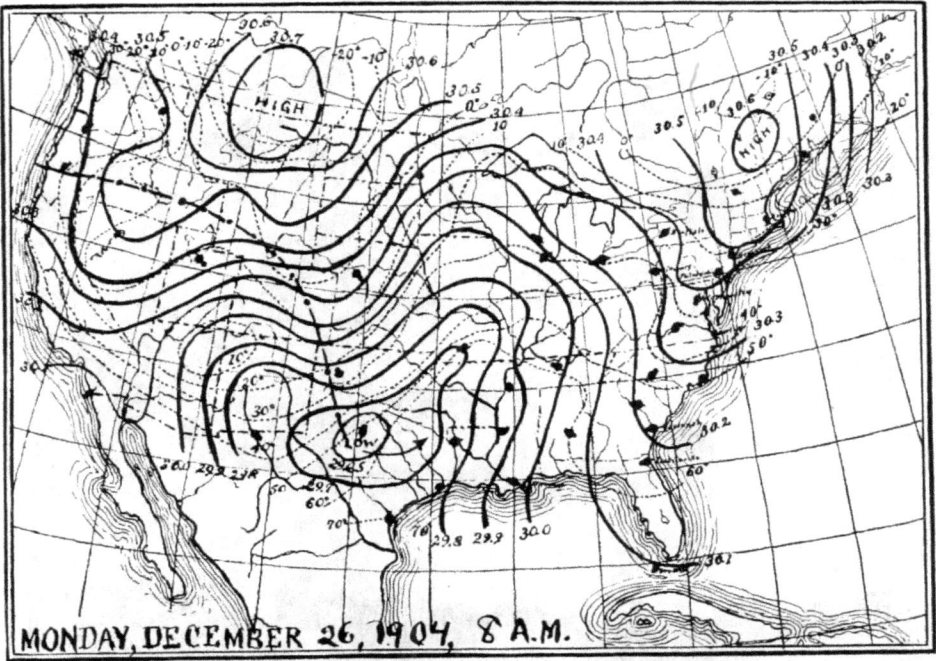

MONDAY, DECEMBER 26, 1904, 8 A.M.

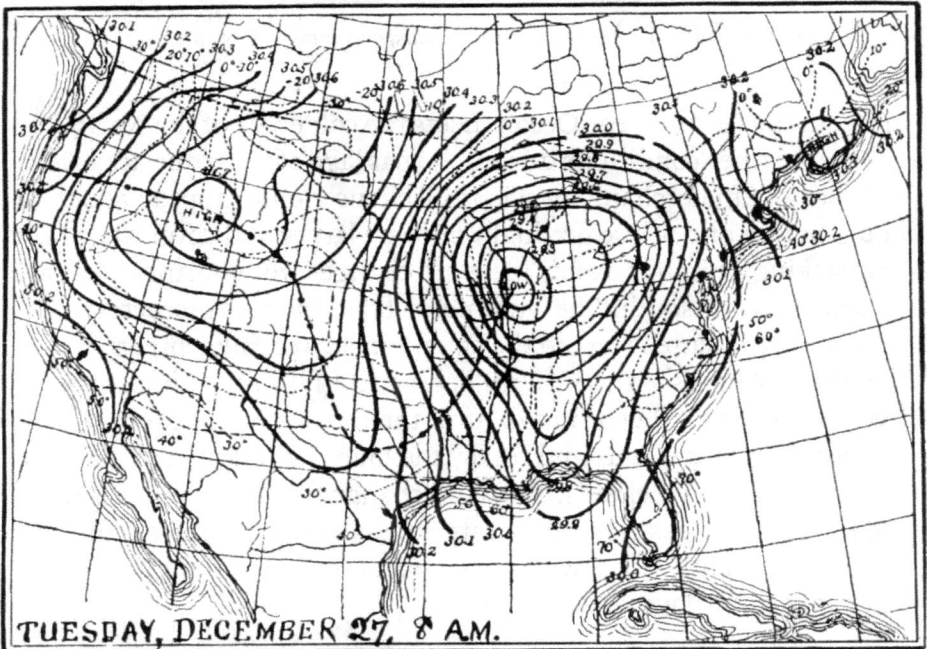

TUESDAY, DECEMBER 27, 8 A.M.

of an area of low pressure for four consecutive days.
is indicated by the line of dots and dashes on the later maps.

A sky filled with many types of cirrus clouds, accompanied by cirrocumulus

air from the Gulf and the Ocean meets the colder air of the North.

The weather conditions during the passage of a cyclone are briefly as follows: Small, changing wisps of cirrus clouds appear about 24 hours before rain; these gradually become larger and cover the whole sky, making a nimbus cloud. The wind changes from northeast to east or southeast to south. The barometer falls; the thermometer rises, that is, air pressure is less to the square inch, and the temperature of the atmosphere is warmer. Rain begins and falls for a time, varying from an hour to a day or more. After the rain there appear breaks in the great nimbus clouds and finally the blue sky conquers until there are only a few or no clouds. The wind changes to southwest and west; the barometer rises, the temperature falls. The rain ceases, the sun shines out brightly. The low has passed and the high is approaching to last about three days.

LESSON

Leading thought— Weather maps are made with great care by the Weather Bureau experts. Each map is the result of many telegraphic communications from all parts of the country. Every intelligent person should be able to understand the weather maps.

Method— Get several weather maps of the nearest Weather Bureau Station. They should be maps for successive days, and there should be enough so that each pupil can have three maps, showing the weather conditions for three successive days.

Observations—

1. Take the map of the earliest date of the three. Where was your map used? What is its date? How many kinds of lines are there on your map? Are there explanatory notes on the lower left-hand corner of your map? Explain what the continuous lines mean. Find an isobar of 30; to what does this figure refer? Find all the towns on your map where the barometer stands at 30 inches. Is there more than one isobar on your map where the barometer stands at 30?

2. Where is the greatest air pressure on your map? How high does the barometer stand there? How are the isobars arranged with reference to this region? What word is printed in the center of this series of isobars?

3. What do the arrows indicate? What do the circles attached to the arrows indicate?

4. In general, what is the direction of the winds with reference to this high center?

5. Is the air rising or sinking at the center of this area? If the wind is blowing in all directions from a center marked high, what sort of weather must the places just east of the high be having? Do the arrows with their circles indicate this?

6. Find a center marked low. How high does the barometer stand there? Does the air pressure increase or diminish away from the center marked low, as indicated by the isobars? Do the winds blow toward this center or away from it?

7. What must the weather in the region just east of the low be? Why? Do the arrows and circles indicate this?

8. Is there a shaded area on your map? If so, what does this show?

9. Compare the map of the next date with the one you have just studied. Are the highs and lows in just the same position that they were the day before? Where are the centers high and low now? In what directions have they moved?

10. Look at the third map and compare the three maps. Where do the high and low centers seem to have originated? How long does it take a high or low to cross the United States? How far north and south does a high or low, with all its isobars, extend?

11. What do the dotted lines on your map mean? Do they follow exactly the isobars?

12. What is the greatest isotherm on your map? Through or near what towns does it pass?

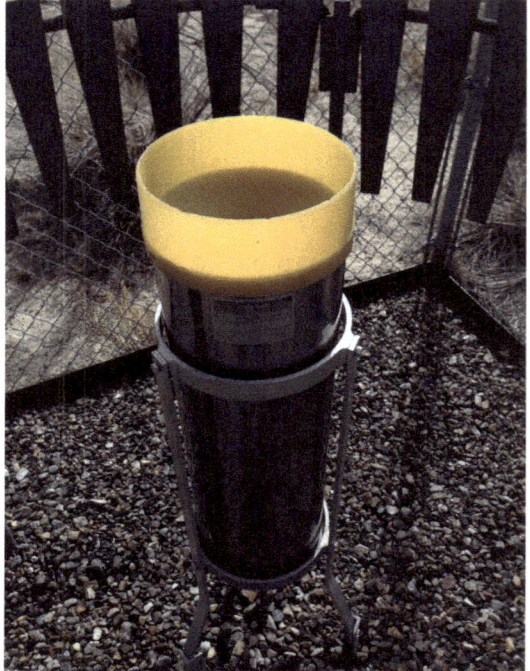

A standard NOAA (National Oceanic and Atmospheric Administration) rain gauge

13. Do the regions of high air pressure have the highest temperature or the lowest? Do high temperatures accompany low pressures? Why?

14. What is the condition of the sky just east of a low center? What is its condition just west of low?

15. If the isobars are near together in a low, it means that the wind is moving rather fast and that there will be a well marked storm. Look at the column giving wind velocity. Was the wind blowing toward the center of the low on the map? If so, does that mean it is coming fast or slow? How does this fact correspond with the indications shown by the distance between the isobars?

16. Describe the weather accompanying the approach and passage of a low in the region where your town is situated. What sort of clouds

would you have, what winds, what change of the barometer and thermometer?

HOW TO FIND THE GENERAL DIRECTION AND AVERAGE RATE OF MOTION OF HIGHS AND LOWS

Observations—

1. On the first map of the series of three given, put an X in red pencil or crayon at the center of the high and a blue one at the center of the low; or if you do not have the colored pencils, use some other distinguishing marks for the two. If there are two highs and two lows use a different mark for each one.

2. Mark the position of each center on this map for the following day with the same mark that you first used for that area. Do this for each of the highs and lows until it leaves the map or until your maps have been used. All the marks of one kind can be joined by a line, using a red line for the red marks and a blue line for the blue marks.

3. What do you find to be the general direction of the movement of the highs and lows?

4. Examine the scale marked statute miles at the bottom of the map. How many miles are represented by one inch on the scale?

5. With your ruler find out how many miles one area of high or low has moved in twenty-four hours; in three days. Divide the distance which the area has moved in three days by three and this will give the average velocity for one day.

6. In the same way find the average velocity of each of the areas on your map for three days and write down all your answers. From all your results find the average weekly velocity; that is, how many miles per hour and the general direction which has characterized the movement of the high and low areas.

Supplementary reading— The Wonderbook of the Atmosphere, Houston, Chapters XIV-XXIII.

HOW TO KEEP A DAILY WEATHER MAP

THE pupils should keep a daily weather map record for at least six months. The observations should be made twice each day and always at the same hours. While it would be better if these records could be made at 8 o'clock in the morning and again at 8 o'clock in the evening, this is hardly practicable and they should, therefore, be made at 9 o'clock and at 4. The accompanying chart may be drawn enlarged. Sheets of manila paper are often used, so that one chart may cover the observations for a month.

Few schools are able to have a working barometer, but observations of temperature and sky should be made in every school. Almost any boy can make a weather vane, which should be placed on a high building or tree where the wind will not be deflected from its true direction when striking it. A thermometer should be placed on the north side of a post and on a level with the eyes; it should not be hung to a building, as the temperature of the building might affect it.

The direction of the wind and the cloudiness of the day may be indicated on the chart, as it is on the weather maps, by a circle attached to an arrow which points in the direction in which the wind is blowing.

References— *Elementary Meteorology*, Waldo, American Book Co., $1.50; *Elementary Meteorology*, Davis, Ginn and Co., $2.50; *Bulletins from the United States Weather Bureau*, Washington, D. C.

CHART FOR SCHOOL WEATHER-RECORDS.

Date	Hour	Temp.	Barometer	Direction of wind	Cloudiness. Fogs.	Dew or Frost	Rain or Snow	Remarks.
Weekly Summary								

The Story of the Stars

"Why did not somebody teach me the constellations and make me at home in the starry heavens, which are always overhead, and which I don't half know to this day."

—THOMAS CARLYLE

FOR many reasons aside from the mere knowledge acquired, children should be taught to know something of the stars. It is an investment for future years; the stars are a constant reminder to us of the thousands of worlds outside our own, and looking at them intelligently, lifts us out of ourselves in wonder and admiration for the infinity of the universe, and serves to make our own cares and trials seem trivial. The author has not a wide knowledge of the stars; a dozen constellations were taught to her as a little child by her mother, who loved the sky as well as the earth; but perhaps nothing she has ever learned has been to her such a constant source of satisfaction and pleasure as this ability to call a few stars by the names they have borne since the men of ancient times first mapped the heavens. It has given her a sense of friendliness with the night sky, that can only be understood by those who have had a similar experience.

There are three ways in which the mysteries of the skies are made plain to us: First, by the telescope; second, by geometry, trigonometry and calculations—a proof that mathematics is even more of a heavenly than an earthly science; and third, by the use of the spectroscope, which can only be understood after we study physics. It is an instrument which tells us, by analyzing the light of the stars, what chemical elements compose them; and also, by the means of the light, it estimates the rate at which the stars are moving and the direction of their motion.

Thus, we have learned many things about the stars; we know that every shining star is a great blazing sun, and there is no reason to doubt that many of these suns have worlds, like the earth, spinning around them although, of course, so far away as to be invisible to us; for our world could not be seen at all from even the nearest star. We also know that many of the stars which seem single to us are really double—made up of two vast suns swinging around a common center; and although they may be millions of miles apart, they are so far away that they seem to us as one star. The telescope reveals many of these double stars and shows that they circle around their orbits in various periods of time, the most rapid making the circle in five years, another in sixteen years, another in forty-six years; while there is at least one lazy pair which seems to require fully sixteen hundred years to complete their circle. And the spectroscope has revealed to us that many of the stars which seem single through the largest telescope are really double, and some of these great suns race around each other in the period of a few hours, which is a rate of speed we could hardly imagine.

Astronomers have been able to measure the distance from us to many of the stars, but when this distance is expressed in miles it is too much for us to grasp. Thus, they have come to measure heavenly distance in terms of the rate at which light travels, which is 186,400 miles per second or about six trillions of miles per year; this distance is called a light-year. Light reaches us from the sun in about eight minutes, but it takes more than four years for a ray to reach us from the nearest star. It adds new interest to the Pole-star to know that the light which reaches our eyes left that star almost half a century ago, and that the

light we get from the Pleiades may have started on its journey before America was discovered. Most of the stars are so far away that we cannot measure the distance.

Although the stars seem always to be in the same places, they are all moving through space just as our sun and its family are doing; but the stars are so far away that, although one may move a million miles a day, it would require many years of observation for us to detect that it moved at all. We know the rate at which some of the stars are moving but have no idea of their goal; nor do we have any idea where our sun is dragging us at the rate of nearly 800 miles per minute. It is thought that our sun and the other suns are whirling around some greater center or centers; but if so, the orbits are so many trillions of miles across that the suns all seem to be going somewhere in a straight line, each attending strictly to its own business.

Through the spectroscope we know something of the life of stars; we know that when they are young they are composed of thin gases and shine white or blue; and as they grow older, they become more solid and shine yellow, like our sun; and when older still, they grow red and are yet more condensed, like Betelgeuse in Orion, which is an aged sun and which will, in time, grow cold and dark and invisible to us. The spectroscope reveals many dark stars whirling through space—vast, dead suns with their fires extinguished, never to be lighted again unless, in its swift course, one of them should hurl itself against another star with a fearful force which shall shatter it into gaseous atoms, and these be thrown into fierce spiral whirlpools, from which it shall again be fashioned into a white-hot sun and become a star in our sky.

The scientists are coming to understand a little of how the stars are made; for scattered through the skies are masses of misty light, called nebulae, which means clouds; nebulae are vast gaseous bodies composed of the stuff of which stars are made. Each nebula keeps its own special place in the heavens—just like a star, and is moving through space—like a star. The spectroscope shows that many of these shining, misty masses are made up of glowing gases, largely hydrogen; and many are disk-shaped, twisted into a spiral. There are grounds for believing that these spiral nebulae are stars in the process of forming. Nebulae are mostly telescopic, although two or three may be detected by the keen, unaided eye as a little blur of light, like that surrounding the third star of Orion's sword. There are eight thousand or more nebulae which have been discovered and mapped. Some idea of their tremendous size is given by Ball when speaking of the ring nebula of Lyra, which we cannot see with the naked eye, and yet if a railroad train started to cross its diameter at the middle, and went at the rate of a mile a minute, one thousand years would not complete the journey.

The number of stars that may be seen with the unaided eye, if one were to travel from the southern to the northern polar region, would be between six and seven thousand; but it would require very keen eyes to see two thousand at one time. With the help of the telescope, about eight hundred thousand stars have been discovered, classified and catalogued, while photography of the skies reveals millions. It is thought that the new international photographic chart, which shall cover all the space seen from our globe, may show thirty millions of stars. The Milky Way or Galaxy, that great, white band across the heavens, is made up of stars which are so far away that we cannot see them, but see only their diffused light. It is well called a "River of

Stars" flowing in a circle around our whole solar system; and, except during the spring months, one-half of it may be seen directly above us while the other half is hidden below us. The place of the Milky Way in the heavens seems fixed and eternal; any star within its borders is always seen at the same point. When the Northern Cross lifts itself toward the zenith we are able to see that, near that constellation, the star river divides into three streams with long, blue islands between.

Reference books— There are a large number of excellent text-books and popular books on astronomy. The following are a few which I have used most often: *Astronomy for Everybody*, Newcomb; *Todd's New Astronomy*; *The Friendly Stars*, Martin; *Starland*, Ball; *The Stars Through an Opera Glass*, Serviss; *Other Suns than Ours*, Proctor; *Other Worlds than Ours*, Proctor.

For children— *Earth and Sky*, Holden; *Stories of Starland*, Proctor; *The Children's Book of Stars*, G. E. Mitton; *Storyland of the Stars*, Pratt; *Stars in Song and Legend*, Porter; *The Planisphere*, Thos. Whittaker.

The big dippers bowl as seen from the International Space Station

How To Begin Star Study

THE POLE-STAR AND THE DIPPERS

THE way to begin star study is to learn to know the Big Dipper, and through its pointers to distinguish the Pole-star; for whenever we try to find any star we have to find the Big Dipper and Pole-star first so as to have some fixed point to start from. There are four stars in the bowl of the Big Dipper and three in the curved handle. A line drawn through the outer two stars of the bowl, if extended, will touch the North Star, or Pole-star. It is very important for us to know the Pole-star, because the northern end of the earth's axis is directed toward it, and it is therefore situated in the heavens almost directly above our North Pole. For those of us who live in the northern Hemisphere, the North Star never sets, but is always in our sky. Of course, the North Star has nothing to do with the axis of our earth any more than the figure on the blackboard has to do with the pointer; it simply happens to lie in the direction toward which the northern end of the earth's axis points. In the southern skies, there is no convenient star which lies directly above the South Pole, so there is no South Pole-star. It

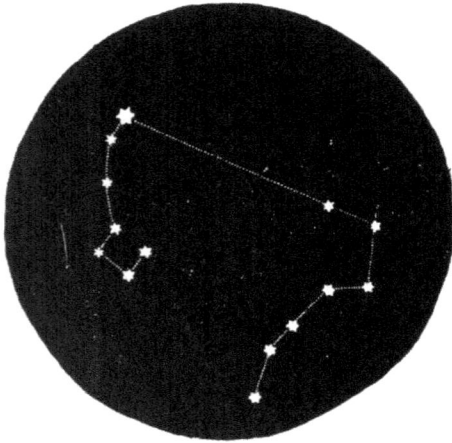

The Pole-star and the Big and Little Dippers.

is also a coincidence that the needle of the mariner's compass points toward the North Star; the earth being a large magnet exercises its influence on all substances which can be magnetized, and since the poles of our great earth-magnet are nearly in line with the poles of the earth's axis, the magnetic needle naturally points north and south, and the North Star chances to be nearly in the direction toward which the northern end of the compass needle points.

The Pole-star cannot be seen from the southern hemisphere; but if we should start from Florida, on a journey toward Baffin's Bay, we should discover that each night this star would seem higher in the sky. And if we should succeed in reaching the North Pole, we would find the Pole-star directly over our heads, and what a wonderful sight the stars would be from this point! For none of the stars which we could see would rise or set, but would move around us in circles parallel to the horizon.

The Big Dipper points towards the Pole-star, and to us seems to revolve around it every twenty-four hours but, of course, this appearance is caused by the fact that we ourselves are revolving from west to east. Therefore, the stars seem to revolve from west to east under the Pole-star and from east to west above it, or in exactly the opposite direction in which the hands of a clock turn. Owing to the movement of the earth in its orbit, the Big Dipper and all the other stars arrive at a certain point in our sky four minutes earlier each day or about two hours earlier each month; thus, the Big Dipper is east of the Pole-star with handle down in the evenings of January, while at the same time of night in July, it is west of the Pole-star with the handle up. But the time of year that a certain star reaches a certain point is so invariable, that if we know star time, or sidereal time as it is called, we can tell just what hour of the night it is when a star passes this point. Thus, the Big Dipper and the other polar constellations are the night clock of the sailors of the northern hemisphere; for though this great polar clock has its hands moving around the wrong way, it gains time with such regularity that anyone who understands is able to compute exact time by it.

The Little Dipper lies much nearer the Pole-star than does the Big Dipper; in fact, the Pole-star itself is the end of the handle of the Little Dipper. Besides the Pole-star, there are two more stars in the handle of the Little Dipper, and of the four stars which make the bowl, the two that form the outer edge are much the brighter. The bowl of the Little Dipper is above or below the Pole-star according to the hour of the evening, or the night of the year, for it apparently revolves about the Pole-star as does the Big Dipper. The two Dippers open toward each other, and some one said "they pour into each other."

The Big Dipper is a part of a constellation called *Ursa Major*, the Great Bear; and the Little Dipper is the Little Bear, the handle of the dipper being the bear's tail.

There is an ancient myth telling the story of the Big and Little Bears: A beautiful mother called Callisto had a little son whom she named Arcas. Callisto was so beautiful that she awakened the anger of Juno, who changed her to a bear; and when her son grew up he became a hunter, and one day would have killed his transformed mother; but Jupiter seeing the danger of this crime caught the two up into the heavens, and set them there as shining stars. But Juno was still vindictive, so she wrought a spell which never allowed these stars to rise and set like other stars, but kept them always moving around and around.

References— *The Friendly Stars* by Martin is a most delightful book and at the same time gives explicit directions for finding the stars and much interesting information concerning them. The planisphere is a little chart with a mechanical device which enables us to find what stars are in sight every night of the year, or at any time of night. It is published by Thos. Whittaker, Bible House, New York.

A 45minute exposure of the stars around Polaris, the Pole-Star

LESSON

Leading thought— The North Star or Pole-star may always be found by the stars known as the pointers in the Big Dipper; the stars of the Big Dipper seem to revolve around the Pole-star once in twenty-four hours.

Method— The time to begin these observations is when the moon is in its last quarter, so that the moonlight will not make pale the stars in early evening. Draw upon the blackboard, from the chart shown, the Big Dipper and the Pole-star, with a line extending through the pointers. Say to the pupils that this Big Dipper is above or below or at one side of the Pole-star, and that you wish them to observe for themselves where it is and tell you about it the next day. After they surely know the Big Dipper, ask the following questions:

Observations—

1. Can you find the Big Dipper among the stars?

2. Is it in the north, south, east or west?

3. Which stars are the "pointers" in the Dipper, and why are they called so?

4. Make a drawing showing how you can always find the Pole-star, if you can see the Big Dipper.

5. How many stars make the bowl of the Dipper?

6. How many stars in the handle?

7. Is the handle straight or is it curved?

8. Does the Big Dipper open toward the Pole-star, or away from it?

9. Is it above or below the Pole-star at eight o'clock in the evening, or at the right or the left of it?

10. Does the Big Dipper remain in the same direction from the Pole-star all night? Look at it at seven o'clock and again at nine o'clock and see if it has changed position.

11. Do you think it moves around the Pole-star once every twenty-four hours? In which direction? How could you tell the time of night by the Big Dipper and the Pole-star?

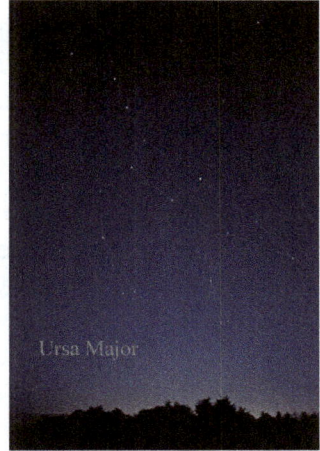

TILL CREDNER (CC BY-SA 3.0)
The constellation Ursa Major as it can be seen by the unaided eye.

12. Does the Big Dipper ever rise and set?

13. The Big Dipper is also called the Great Bear. Can you find the stars which make the bear's head and front legs?

After the pupils surely know the Big Dipper and Pole-star draw the complete diagram upon the board to show the Little Dipper and where it may be found, and call attention to the fact that the end of the Little Dipper's handle is the Pole-star itself and that its bowl is not flaring, like that of the Big Dipper and that the two pour into each other. Let the pupils find the Little Dipper in the sky for themselves and ask the following questions:

Observations—

14. Is the Little Dipper nearer or farther from the Pole-star than the Big Dipper?

15. How many stars in the handle of the Little Dipper?

16. How many stars make the bowl of the Little Dipper? Which of these stars are the brightest? Is the bowl of the Little Dipper above or below the Pole-star?

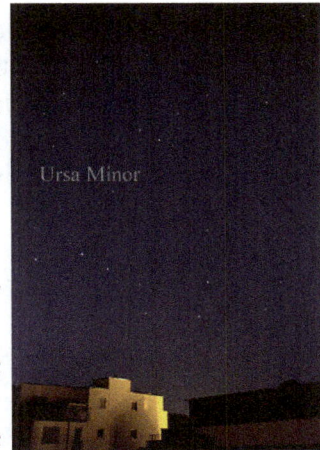

TILL CREDNER (CC BY-SA 3.0)
The constellation Ursa Minor as it can be seen by the naked eye (with connections and label added).

17. Does the Little Dipper extend in the same direction in relation to the Pole-star all night?

18. Make observations on the relation to each other of the two Dippers at eight o'clock in the evening of January, February, March and April.

After the above lessons are well learned, give the following questions, and try to have the pupils answer by thinking:

Questions about Polaris (the North Star) for the pupils to think about and answer:

19. How many names has the Pole-star? Can the Pole-star be seen from the southern hemisphere? If not, why not?

20. If you should start from southern Florida and travel straight north, how would the Pole-star seem to change position each succeeding night?

21. If you could stand at the North Pole, where would the Pole-star seem to be?

22. If you were at the North Pole, would any of the stars rise and set? In what direction would the stars seem to move and why?

23. How does the North Star help the sailors to navigate the seas and why?

24. How do astronomers reckon distances between us and the stars? What is a light-year?

Topics for English lesson— (a) What a star is. (b) What a constellation is. (c) How the stars and constellations received their names in ancient times. In ancient times the Big and Little Dippers were named the Big and Little Bears, and that is their Latin name to this day. Write a story about what the ancient Greeks told about these Bears and how they came to be in the sky.

Supplementary reading— *Stories of Starland*, Proctor, pp. 117-121; *Storyland of the Stars*, Pratt, p. 75; *Child's Study of the Classics*, p. 33.

Cassiopeia in the night sky

Cassiopeia's Chair, Cepheus, and the Dragon

TEACHER'S STORY

THERE are other constellations besides the two Dippers, which never rise and set in this latitude, because they are so near to the Pole-star that, when revolving around it, they do not fall below the horizon. There is one very brilliant star, called Capella, which almost belongs to the polar constellations but not quite, for it is far enough away from Polaris to dip below the horizon for four hours of the twenty-four.

Queen Cassiopeia's Chair is on the opposite side of the Pole-star from the Big Dipper and at about equal distance from it. It consists of five brilliant stars that form a W with the top toward Polaris, one-half of the W being wider than the other. There is a less brilliant sixth star which finishes out half of the W into a chair seat, making of the figure a very uneasy looking throne for a poor queen to sit upon.

King Cepheus is Queen Cassiopeia's husband, and he sits with one foot on the Pole-star quite near to his royal spouse. His constellation is

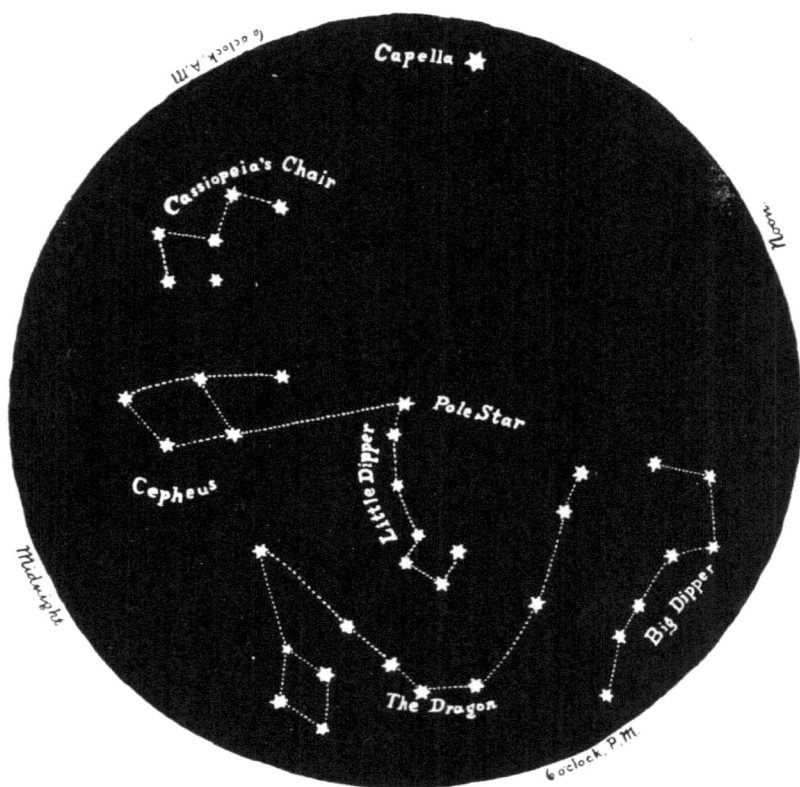

The polar constellations as they appear at about 8 o'clock January 20, the Dragon being south of the Pole-star. By revolving this chart as indicated, the positions of the stars is shown for 6 p.m., midnight, 6 a.m., and noon of January 20th.

marked by five stars, four of which form a lozenge, and a line connecting the two stars on the side of the lozenge farthest from Cassiopeia, if extended, will reach the Pole-star as surely as a line from the Big Dipper pointers. Cepheus is not such a shining light in the heavens as is his wife, for his stars are not so brilliant. Perhaps this is because he was only incidentally put in the skies. He was merely the consort of Queen Cassiopeia, who being a vain and jealous lady boasted that she and her daughter, Andromeda, were far more beautiful than any other goddesses that ever were, and thus incurred the wrath of Juno and Jupiter who set the whole family "sky high" and quite out of the way, a punishment which must have had its compensations since they are where the world of men may look at and admire them for all ages.

Lying between the Big and Little Dippers and extending beyond the latter is a straggling line of stars, which, if connected by a line, make a very satisfactory dragon. Nine stars form his body, three his head, the two brighter ones being the eyes.

LESSON

Leading thought— To learn to know and to map the constellations which are so near the Pole-star that they never rise or set in our latitude, but seem to swing around the North Star once in twenty-four hours.

Method— Place on the blackboard the diagram given showing the Pole-star, Big and Little Dippers and Cassiopeia's Chair, and ask for observations and sketches showing their position in the skies the following evening. After the pupils have observed the Chair and know it, add to your diagram, first Cepheus and then the Dragon. After you are sure the pupils know these constellations, give the following lesson. The observations should be made early and late in the same evening and at different times of the month, so that pupils will in every case note the apparent movement of these stars around Polaris.

Observations—

1. How many stars form Cassiopeia's Chair? Make a drawing showing them and their relation to the Pole-star.

2. Is the Queen's Chair on the same side of the Pole-star as the Big Dipper? Is the top or the bottom of the "W" which forms Cassiopeia's Chair turned toward the Pole-star?

3. Does Cassiopeia's Chair move around the Pole-star, like the Big Dipper?

4. How many stars mark the constellation of Cepheus?

5. Make a sketch of these stars and show the two which are pointers toward the North Star.

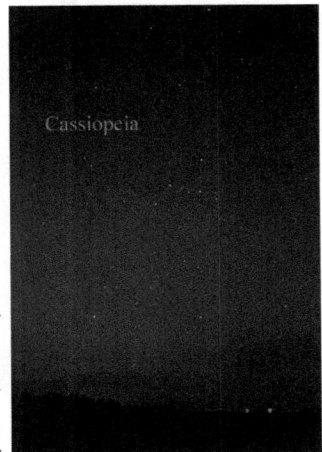

TILL CREDNER (CC BY-SA 3.0)
Cassiopeia with stars joined

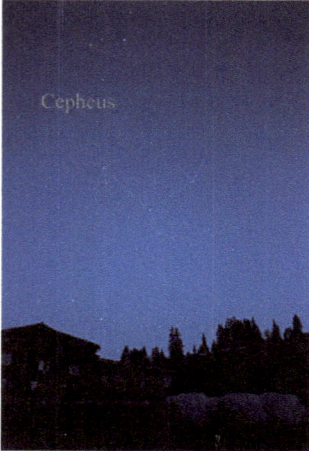

Cepheus with stars joined

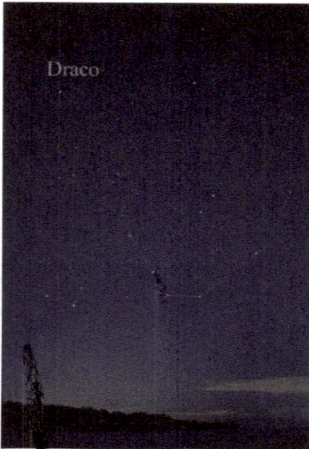

Draco (the dragon) with stars joined

6. Does Cepheus also move around the Pole-star, and in which direction?

7. Describe where the Dragon lies, and where are his tail and his head in relation to the two Dippers. Make a sketch of the Dragon.

8. Why do all the popular constellations seem to move around the Pole-star every twenty-four hours, and why do they seem to go in a direction opposite the movement of the hands of a clock? What do we mean by "Polar constellations"?

Topics for English Themes— The Story of Queen Cassiopeia, King Cepheus and their daughter, Andromeda; the story of the Dragon.

Supplementary reading— *Storyland of the Stars*, Pratt.

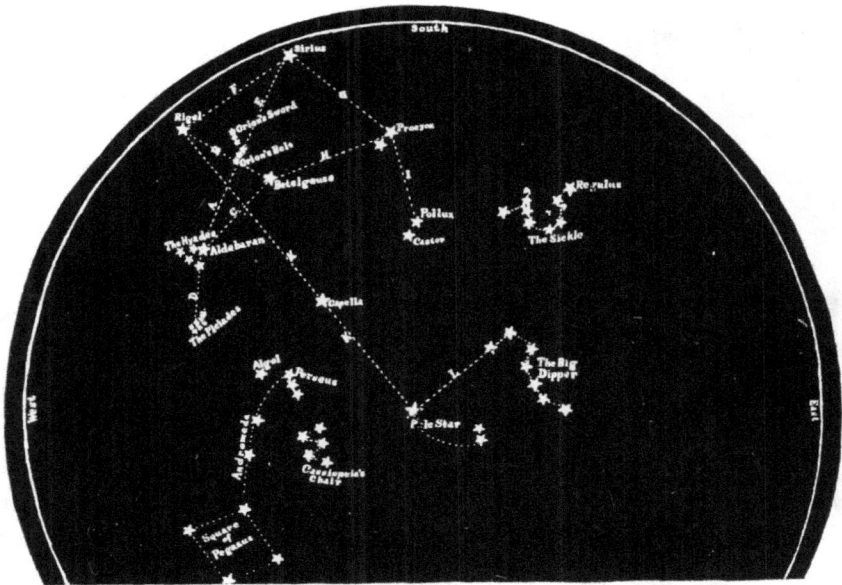

A diagram of the principal stars of winter as seen in early evening late in February.
To use the chart take it in the hands, face the Pole-star and hold the chart above the head so that the side marked east will extend eastward.

The Winter Stars

TEACHER'S STORY

THE natural time for beginning star study is in the autumn when the days are shortening and the early evenings give us opportunity for observation. After the polar constellations are learned, we are then ready for further study in the still earlier evenings of winter, when the clear atmosphere and beautiful blue of the heavens make the stars seem more alive, more sparkling, and more beautiful than at any other period of the year. One of the first lessons should be to instruct the pupils how to draw an imaginary straight line from one star to another, and to perceive the angles which such lines make when they meet at a given star. A rule, or what is just as effective, a postal card or some other piece of stiff paper which shows right-angled corners, is very useful in this work. It should be held between the eyes and the stars which we wish to connect, and thus make us certain of a straight line and a right angle.

Orion, the three large stars in a line forming the belt, the curved line of smaller stars below forming the sword, Betelgeuse above, Rigel below.

To use the chart take it in the hands, face the Pole-star and hold the chart above the head so that the side marked east will extend eastward.

ORION

During the evenings of January, February and March the splendid constellation of Orion (o-ri'on) takes possession of the southern half of the heavens; and so striking is it that we find other stars by referring to it instead of to the Pole-star. Orion is a constellation which almost everyone knows; three stars in a row outline his belt, and a curving line of stars, set obliquely below the belt, outline the sword. Above the belt in the evening sky we can see the splendid red star Betelgeuse (bet-el-gerz) , and below the belt, at about an equal distance, is the white star Rigel (re-jel) . West of the red star above, and east of the white star below, are two fainter stars, and if these four stars are connected by lines, an irregular four-sided figure results, which includes the belt and the sword. In this constellation the ancients saw Orion, the great hunter, with his belt and his sword; Betelgeuse was set like a glowing ruby on his shoulder, and the white star Rigel was set like a spur on his heel. Thus, stood the great hunter in the sky, with his club raised to keep off the plunging bull whose eye is the red Aldebaran (al-deb'a-ran) . And beyond him follows the Great Dog with the bright blue star Sirius (sir'i-us) in his mouth, and the Little Dog branded by the white star Procyon (pro'si-on) . However, our New England ancestors did not see this grand figure in the sky; they called the constellation the Yard-ell or the Ell-yard.

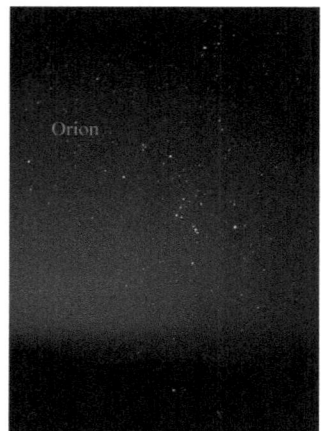

TILL CREDNER (CC BY-SA 3.0)
Orion with stars joined

Deep sky image of Orion

The three beautiful stars which make Orion's belt are all double stars; the belt is just three degrees long and is a good unit for sky measurement. The sword is not merely the three stars which we ordinarily see, but is really a curved line of five stars; and what seems to be the third star from the tip of the sword and which looks hazy, is in fact a great nebula. Through the telescope this nebula seems a splash of light with six beautiful stars within it. Betelgeuse is a brilliant red star, and is the first star in the constellation to appear above the horizon. It is an old, old star and is growing cold, as is shown by its red glow. It glows redder sometimes than at others; it is so far away that we have not been able to measure its distance from us surely, and it is receding from us all the time. About fifteen minutes after Betelgeuse rises, and after the belt and sword are in sight, a white sparkling star appears at the southwest of the belt. This is Rigel, and this star, too, is so far from us that we do not know the distance, and it is also receding.

LESSON

Leading thought— Orion is one of the most beautiful constellations in the heavens. It is especially marked by the three stars which form Orion's belt, and the line of stars below the belt which form the sword.

Method— Place on the blackboard the outline of Orion as given in the diagram. Ask the pupils to make the following observations in the evening and give their report the next day.

Observations—

1. Where is Orion in relation to the Pole-star?

2. How many stars in the belt of Orion? How many stars in the sword? Can you see plainly the third star from the bottom of the sword?

3. Notice above the belt, about three times its length, a bright star; this is Betelgeuse. What is the color of this star? What do we know about the age of a star if it is red?

4. Look below the belt and observe another bright star at about the same distance below that Betelgeuse is above. What is the color of this star? What does its color signify? The name of this star is Rigel.

5. Note that west of the red star above and east of the white star below are two fainter stars. If we connect these four stars by lines we shall make an irregular four-sided figure, fencing in the belt and sword. Sketch this figure with the belt and sword, and write on your diagram the name of the red star above and the white star below and also the name of the constellation.

6. Which star of the constellation rises first in the evening? Which last?

7. Write an English theme on the story of Orion, the great hunter.

Supplementary reading— *Stories of Starland*, Proctor; *The Stars in Song and Legend*, Porter; *Storyland of the Stars*, Pratt.

Hyades and Pleiades

Aldebaran and the Pleiades

TEACHER'S STORY

ALMOST in a line with the belt of Orion, up in the skies northwest from it, is the rosy star Aldebaran. This ruddy star, which is not so red as Betelgeuse, marks the end of the lower arm of a V-shaped constellation composed of this and four other stars. This constellation is the Hyades (*hi'a-dez*). The Hyades is a part of the constellation called by the ancients Taurus, the bull, and is the head of the infuriated animal. Aldebaran is a comparatively near neighbor of ours, since it takes light only thirty-two years to pass from it to us. It gives off about forty-five times as much light as does our sun; it lies in the path traversed by the moon as

The Pleiades, a group of six small stars surrounded by a misty light.

Aldebaran in the V-shaped constellation called the Hyades. This is a part of the constellation Taurus.

it crosses the sky, and is often thus hidden from our view.

Although we are attracted by many bright stars in the winter sky, yet there is a little misty group of stars, which has ever held the human attention enthralled, and of which the poets of all the ages have sung. These stars are called the Pleiades (ple'ya-dees); most eyes can count only six stars in the constellation. There are nine stars large enough to be seen through the telescope, and which have been given names; but sky photography has revealed to us that there are more than three thousand stars in this little group. Perhaps no stars in the heavens give us such a feeling of the infinity of the universe as

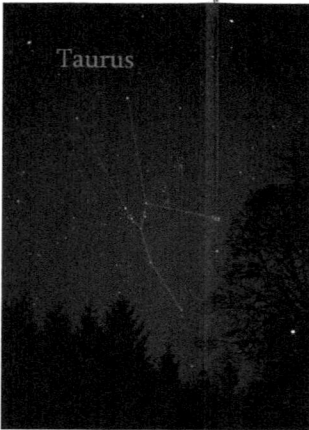

TILL CREDNER (CC BY-SA 3.0)
Aldebaran is the brightest star in the constellation of Taurus

do the Pleiades; for astronomers believe that they form a great star system which is now being evolved from a nebula. The reason for this belief is that these stars seem to be surrounded by a brilliant mist which sometimes seems to be looped from one to another; and, too, the stars are all in the same stage of development and have the same chemical composition, and they are all moving together in the same direction. These stars which look so close together to us are so far apart really that our own sun and all its planets could roll in between them and never be noticed. It would require several years for light to travel from one of these stars in the Pleiades

to another. The Pleiades are so far from us that we cannot estimate the distance, but we know that it takes light several hundred years to reach us from them. There is a mythical story found in literature, that once the unaided eye could see seven instead of six stars in the Pleiades, and much poetic imagining has been developed to account for the "lost Pleiad."

The Pleiades

LESSON

Leading thought— The Pleiades seem to be a little misty group of six stars, but instead there are in it three thousand stars. Half way between the Pleiades and Orion's belt is Aldebaran, an aging ruddy star.

Method— Draw the diagram (p. 131) on the blackboard showing Orion, Aldebaran and the Pleiades, and the lines B, C, D. Give an outline of the observations to be made by the pupils, and let them work out the answers when they have opportunity. Each pupil should prepare a chart of these constellations.

Observations—

1. Imagine a line drawn from Rigel to Betelgeuse and then another line just as long extending to the west of the latter at a little less than a right angle, and it will end in a bright, rosy star, not so red as Betelgeuse.

2. What is the name of this star? Write it on your chart.

3. Can you see the figure V formed by Aldebaran and four fainter stars? Sketch the V and show where in it Aldebaran belongs. This V-shaped constellation is called the Hyades.

4. Imagine a line drawn from Orion's belt to Aldebaran and extend it to not quite an equal length beyond it, and it will end near a "fuzzy little bunch" of stars which are called the Pleiades. Place the Pleiades on your chart.

5. How many stars can you see in the Pleiades?

6. Why are they called the seven sisters?

7. How many stars in the Pleiades which are named, and how many does photography show that there really are in the group?

8. How far apart from each other are the nearest neighbors of the Pleiades?

9. What do the astronomers think about the Pleiades and why do they think this?

The Nebra sky disc, dated circa 1600BC. The cluster of dots in the upper right portion are believed to be the Pleiades

Sirius (bottom) and the constellation Orion (right). The three brightest stars, Sirius, Betelgeuse (top right) and Procyon (top left) form the Winter Triangle

The Two Dog Stars, Sirius and Procyon

TEACHER'S STORY

IF a line from Aldebaran pass through the belt of Orion and is extended about as far on the other side, it will reach the Great Dog Star, following at Orion's heels. This is Sirius (*Sir'-e-us*), the most brilliant of all the stars in our skies, glinting with ever changing colors, sometimes blue, at others rosy or white. It must have been of this star that Browning wrote:

> "All that I know
> Of, a certain star
> Is, it can throw
> (Like the angled spar)
> Now a dart of red,
> Now a dart of blue."

Sirius is a comparatively young star, and is estimated by Proctor to have a diameter of about twelve million miles or fourteen times that of our own sun; it is only eight and one-half light-years away from us and is the most celebrated star in literature. The ancients knew it, the Egyptians worshipped it, Homer sang of it, and it has had its place in the poetry of all ages.

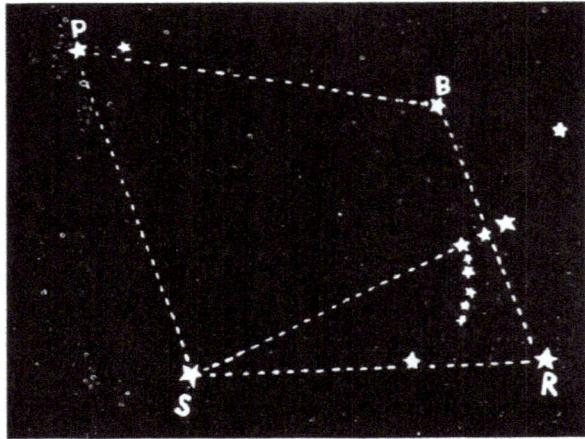
Orion and the Dog Stars.
B. Betelgeuse; R. Rigel; S. Sirius, the Great Dog Star; P. Procyon, the Little Dog Star.

Procyon, (*pro'- se-on*) the Little Dog Star, was so-called perhaps because it trots up the eastern skies a little ahead of the magnificent Great Dog Star; it gives out eight times as much light as our sun, and is only ten light-years away from us. It has a fainter companion about three or four degrees to the northwest of it.

LESSON

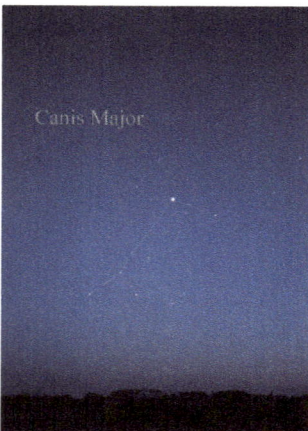
Sirius is the brightest star in the constellation of Canis Major

Leading thought— The Great Dog Star, Sirius, is the most famous of all stars in the literature of the ages. The Two Dog Stars were supposed by the ancients to be following the great hunter, Orion.

Method— Draw upon the board from the chart shown on this page, the constellation of Orion with Sirius and Procyon. Ask the pupils to note that after Orion is well up in the sky a straight line drawn through Orion's belt and dropping down toward the eastern horizon ends in a beautiful white star, which is Sirius. And that if we draw a line from Betelgeuse to Rigel and from Rigel to Sirius and

Sirius in space

then draw lines to complete a quadrangle, we shall find our lines meet at a bright star just a little too far away to make the figure a square, but making it somewhat kite-shaped instead. This is the Little Dog Star, Procyon, and it has a twin star near it. After giving these directions let the children make the following observations.

Observations—

1. How do you find Sirius? Which rises first, Orion or Sirius?

2. What color is Sirius? Judging from its color what stage of development do you think it is in?

3. Try and find out how large Sirius is compared with our sun and how near it is to us.

4. Why is Sirius called the Great Dog Star? Is the Little Dog Star nearer to the North Star than Sirius? Which is the brighter, the Great Dog Star or the Little Dog Star? Can you see any fainter star near Procyon?

5. Why is Procyon called the Little Dog Star?

6. Make a chart showing Orion and the two Dog Stars.

The night sky showing Auriga and the Pleiades. Capella is the brightest star, towards top left

Capella and the Heavenly Twins

TEACHER'S STORY

CAPELLA is nearer to the North Star than any other of the bright stars and it comes very near belonging to the strictly polar constellations, since it falls below the horizon only four hours out of twenty-four. In composition it much resembles our sun, as do all the bright yellow stars; but it is much larger; it gives off one hundred and twenty times as much light as our sun, and it is forty light-years away from us. Capella is always a beautiful feature of the northern skies, being almost in the zenith during the evenings of January and February. It is in a brilliant shield-shaped constellation known as Auriga.

During the winter evenings we see two stars set like glowing eyes almost in the zenith, and in a region of the sky where there are no other bright stars. These twin stars are set just a little closer together than are the pointers of the Big Dipper. To this brilliant pair the ancients gave the names of Castor and Pollux. Pollux is the brighter of the two and is the more southward in situation. Pollux and Castor were two beautiful twin boys who loved each other so much that, after they were dead, they were placed in the skies where they could always be near each other. The twin stars are supposed to exert a benign influence on oceans and seas and are, therefore, beloved by sailors. Although they seem to us so

Capella is the brightest star in the constellation of Auriga (upper left).

near together, they are separated by a space so great that we cannot conceive of it and they are going in opposite directions.

Pollux is a yellow star, and supposed to be in the same stage of development as our sun, while Castor is white and according to star ages is young. When a boy says "By Jimminy," he does not realize that he is using an ancient expletive "By Gemini," which is the Latin name of these twin stars and was a favorite ancient oath, especially of sailors.

LESSON

Capella in the constellation Auriga.

Leading thought— There are, during the evenings of January and February, three brilliant stars almost directly overhead. One of these is Capella, the other two are the Heavenly Twins.

Method— Place on the board the part of the chart (p. 131) showing the Big Dipper, Pole-star, Capella and the Twins. Draw a line, L, from the pointers of the Big Dipper, and extend it to the Pole-star. Draw another line, K, from the Pole-star at right angles to the line L, and on the side away from the Big Dipper's handle, and it will pass through a large, brilliant, yellow star which is Capella. Ask the pupils to imagine similar lines drawn across the sky, when they are making their observations and thus find these stars, and to place them on their charts, making the following observations.

Observations—

1. What color is Capella, and how does its color compare with that of our sun?

2. Is Capella as near to the Pole-star as the Big Dipper? Is it near enough so that it never sets?

3. Can you see the shield-like constellation of which Capella is a part? Do you know the name of this constellation?

Gemini, the heavenly twins, the larger one is Pollux and the other is Castor.

4. How do you find the Heavenly Twins after you have found Capella?

5. Why are these stars called the Heavenly Twins? What is their Latin name? What are the names of the two stars? Are these twins set nearer together than the pointers of the Big Dipper?

6. How can you tell the Heavenly Twins from the Little Dog Star and its companion?

7. Read in the books all that you can find about the Heavenly Twins. Try and find if they are the same age, if they are as near together as they seem, and if they are going in the same direction. What did the ancient sailors think of these twin stars?

TILL CREDNER (CC BY-SA 3.0)

The constellation Gemini as it can be seen with the unaided eye, with added connecting lines.

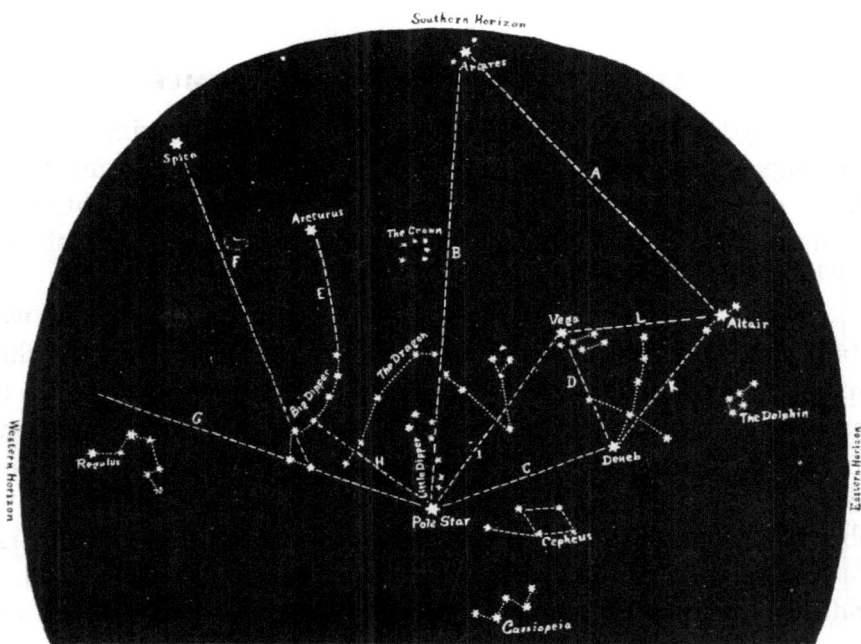

A chart of the brightest stars of summer, showing their positions in early evenings of June.
To find the stars hold the chart above the head and face the north.

The Stars of Summer

TO us, who dwell in a world of change, the stars give the comfort of abidingness; they remain ever the same to our eyes and the teacher should make much of this. When we once come to know a star, we know exactly where to find it in the heavens, wherever we may be. A star which a person knows during childhood will, in later life and in other lands, seem a staunch friend and a bond, drawing him back to his early home and associations.

The summer is an enticing season for making the acquaintance of eight of the fifteen brightest stars visible in northern latitudes. Few midsummer entertainments rival that of lying on one's back on the grass of some open space which commands a wide view of the heavens, and there with a planisphere and an intermittently lighted candle with which to consult it, learn by sight, by name and by heart those brilliant stars which will ever after meet with friendly greeting our uplifted eyes. To teach the children in a true informing way about the stars, the teacher should know them, and nowhere in nature's realm is there a more thought-awakening lesson.

LESSON - THE BRIGHT STARS OF SUMMER

Leading thought— The stars which we see shining during summer evenings are not the same ones that we see during the winter evenings, except those in the polar constellations. There are eight of the brilliant summer stars, which we should be able to distinguish and call by name.

Method— Begin by the middle of May when the Big Dipper is well above the Pole-star in the early evening, and when, therefore, Regulus, Spica, Arcturus and the Crown are high in the sky. The others may be learned in June, although July is the best month for observing them. In teaching the pupils how to find the stars, again instruct them how to draw an imaginary straight line from one star to another and to observe the angles made by such lines connecting three or four stars.

Place upon the blackboard the figures from the chart (page 145), as indicated, leaving each one there until the pupils have observed and learned it. Then erase and place another figure. In each case try to get the pupils interested in what we know about each star, a brief summary of which is given. Note that the observations given in the lessons are for early in the evenings of the last of May, of June, and of early July.

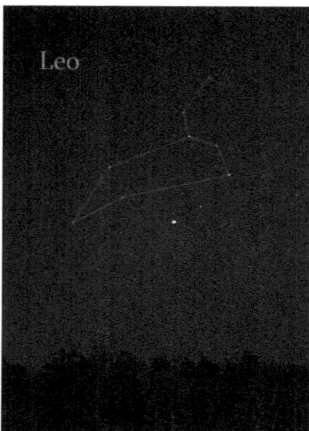

Regulus is the brightest star in the constellation of Leo (right tip)

REGULUS (REG'-U-LUS)

Draw upon the blackboard from the chart (p. 145) the Pole-star, the Big Dipper, the line G and the Sickle shown just below the outer end of the line. Extend the line that passes through the pointers of the Big Dipper to the North Star backward into the western skies; just west of this line lies a constellation called the Sickle, and the stars that form it outline this implement. The Sickle has a jewel at the end of the handle, which is a white and diamond-glittering star called Regulus. It is a great sun giving out one thousand times as much light as our own sun, and this

146

Regulus, the large star in the handle of the sickle.

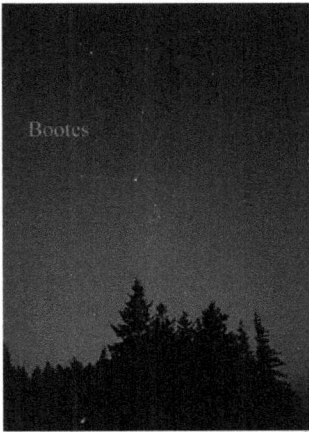

Arcturus is the brightest star in the constellation of Bootes.

light reaches us in about one hundred and sixty years. The Sickle is part of a constellation called the Lion, and from which comes the shower of meteors which we see on the evening of November 13th. Regulus is seen best in Spring.

ARCTURUS (ARK-TU'RUS)

Place on the blackboard the Big Dipper, the Pole-star and the line E, Arcturus and the Crown. Extend the handle of the Big Dipper following its own curve for about twice its length and it will end in a beautiful, yellow star, the only very bright one in that region. It is a thousand times brighter than our own sun, but its light does not reach us for a hundred years after it is given off. Arcturus is supposed to be one of the largest of all the suns, having a diameter of several millions of miles. During the latter part of June and July it is almost overhead in the early evening.

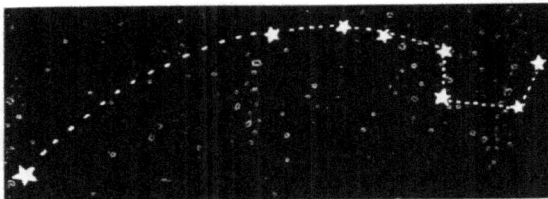

Arcturus and the Big Dipper.

THE CROWN

The Northern Crown.

Between Arcturus and Vega, but much nearer the former, is a circle of smaller stars that is called the Northern Crown, and which because of its form is quite noticeable.

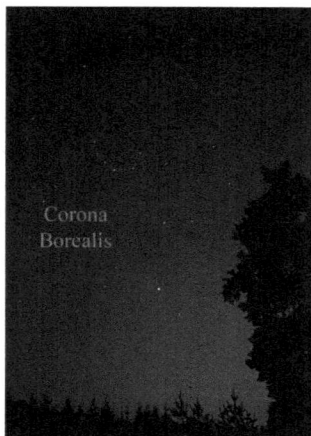

The constellation Corona Borealis as it can be seen by the naked eye

SPICA (SPI'-KA)

Place on the blackboard the Big Dipper, the Pole-star, the line F and Spica. To find Spica draw a line through the star on the outer edge of the top of the bowl of the Big Dipper, through the star at the bottom of the bowl next the handle, and extend this line far over to the southwest, during the evenings of June and July. In August, this star sets at ten o'clock. Spica is a white star, and is the only bright one in that part of the sky. It is so far away from us that the distance has never been measured. Spica is in the constellation called the Virgin.

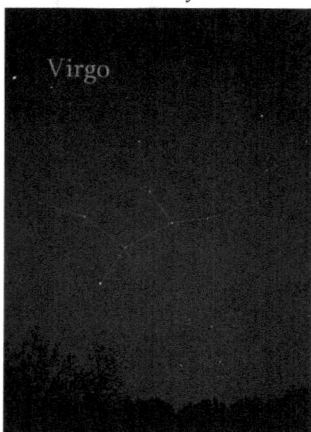

Spica is the brightest star in the constellation of Virgo (lower left).

VEGA (VEE'-GA)

Place on the blackboard the Pole-star, the Big Dipper, the lines H and I and Vega with her five attendant stars, as shown in the chart. Teach that these stars are the chief ones in the constellation called *The Lyre*. To find Vega, draw a line from the Pole-star to the star in the Big Dipper which

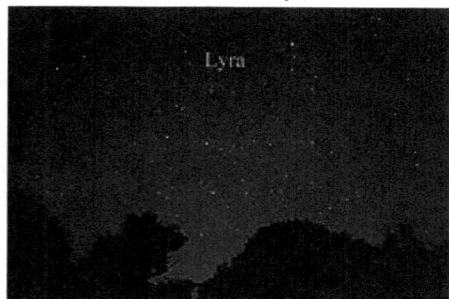

Vega is the brightest star in the constellation of Lyra

Vega and her train of five stars.

joins the bowl to the handle. Then draw a line at right angles to this (see chart lines H, I) and extend the line I a little farther from the North-star than is the end star of the Dipper handle; this line will reach a bright star, bluish in color, which can always be identified by four smaller attendant stars which lie near it and outline a parallelogram with slanting ends. Vega is the most brilliant summer star that we see in the northern hemisphere. It is a very large sun, giving out ninety times as much light as our sun; it is so far away that it requires twenty-nine years for a ray of light to reach us from it. Vega's chief interest for us, aside from its beauty, is that toward it our sun and all its planets, including our earth, are moving at the rate of thirteen miles per second.

ANTARES (AN-TA'-REES)

Add to the last diagram on the blackboard the line E, Arcturus, the line B and Antares. To find this star, draw a line half way between Arcturus and Vega from the Pole-star straight across the sky to the south, and just above the southern horizon it will point to the glowing star, Antares, in the constellation of the Scorpion. Also a line drawn at right angles to the line connecting Altair with its companions and extending toward the south will reach Antares. Late June and July about ten o'clock in the evening is the best for viewing this beautiful star. An interesting thing about Antares

Antares (bright star upper left) as seen from the ground

Antares, a brilliant star in the southern skies.

is that, although it is red, it has, whirling around it, a companion star which is bright green.

DENEB, OR ARIDED (DEN'-EB; A'-RI-DED)

Erase from the last diagram Antares and the line B. Add to it the lines C and D making a right angle at Deneb; and the Cross—the head of which is Deneb, the foot ending near the letter on line L. This star is at the head of the Northern Cross, which is a very shaky looking cross and appears upside down in the eastern skies during the evenings of June and July. Deneb is white in color and is a very large sun, because it seems to us a bright star although it is so far away from us that the distance has never been surely measured; but it has been estimated that a ray of light would need at least three hundred and twenty-five years to reach us from Deneb. It and the cross are a part of the constellation of Cygnus, or the Swan.

The Northern Cross, in the constellation of the Swan.

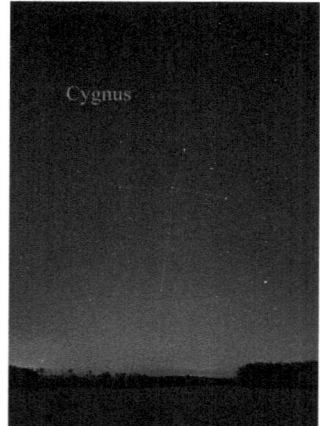
TILL CREDNER (CC BY-SA 3.0)
Deneb is the brighest star in the constellation of Cygnus (top)

ALTAIR

Add to the last diagram on the board the lines L, K, Altair and its two attendant stars and the Dolphin. Emphasize the fact that Altair marks the constellation of Aquila, or the Eagle. This beautiful star is easily distinguished because of its small companions, one on each side, all three in a line. The three belong to a constellation called the Eagle, and may be seen in early evening from June to December.

Altair in the constellation of the Eagle.

Altair is the brightest star in the constellation Aquila

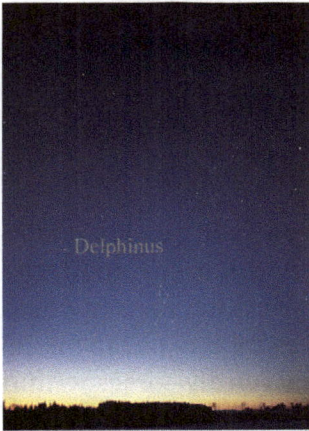

The constellation Delphinus as it can be seen by the naked eye. The main asterism in Delphinus is Job's Coffin

Altair, Deneb and Vega form a triangle with the most acute angle at Altair. (See chart L, K.) Just northeast of Altair is a little diamond-shaped cluster of stars called the Dolphin, which is a good name for it, since it looks like a dolphin, the fifth star forming the tail. It is also called Job's Coffin, but the reason for this is uncertain, unless Job's trials extended to a coffin which could not possibly fit him. If the line C on the chart drawn from the Pole-star to Deneb be extended, it will touch the Dolphin. Altair is always low in the sky; it is a great sun giving off nearly ten times as much light as our own sun; light reaches us from it in fifteen years.

The Dolphin or Job's Coffin.

The Sun

IF, only once in a century, there came to us from our great sun, light and heat, bringing the power to awaken dormant life, to lift the plant from the seed and clothe the earth with verdure, then it would indeed be a miracle. But the sun by shining every day cheapens its miracles in the eyes of the thoughtless. While it hardly comes within the province of the nature-study teacher to make a careful study of the sun, yet she may surely stimulate in her pupils a desire to know something of this great luminous center of our system.

Our sun is a great shining globe about one hundred and ten times as thick through as the earth, and more than a million times as large. If we look at the sun in a clear sky, it is so brilliant that it hurts our eyes. Thus, it is better to look at it through a smoked glass, or when the atmosphere is very hazy. If we should see the sun through a telescope, we should find that its surface is not one great glare of light but is mottled, looking like a plate of rice soup, and at times there are dark spots to be seen upon its surface. Some of these spots are so large that during very "smoky weather" we can see them with the naked eye. In September, 1908, a sun-spot was plainly visible; it was ten thousand miles across, and our whole world could have been dropped into it with a thousand miles to spare all around it. We do not know the cause of these sun-spots, but we know they appear in greater numbers in certain regions of the sun, above and below the equator. And since each sun-spot retains its place on the surface of the sun, just as a hole dug in the surface of our earth would retain its place, we have been able to tell by the apparent movement of these spots how rapidly and in which direction the sun is turning on its axis; it revolves once in about twenty-six days and, since the sun is so much larger than our earth, a spot on the equator travels at a rate of more than a mile a second. There is a queer thing about the outside surface of the sun—the equator rotates more rapidly than the parts lying nearer the poles; this shows that the sun is a gaseous or liquid body, for if it were solid, like our earth, all its parts would have to rotate at the same rate. At periods of eleven years the greatest number of spots appear upon the sun.

Another interesting feature of the sun is the tremendous explosion of hydrogen gas mixed with the vapors of calcium and magnesium, which shoot out flames from twenty-five thousand to three hundred thousand miles high, at a rate of speed two hundred times as swift as a rifle bullet travels. Think what fireworks one might see from the sun's surface all the time! One would not need to wait until the Fourth of July for fireworks there. These great, explosive flames can be seen by the telescope when the moon eclipses the sun, and they have been analyzed by means of the spectroscope. Besides these magnificent explosions, there is surrounding the sun a glow which is brighter near the sun's surface and paler at the edges; it is a magnificent solar halo, some of its streamers being millions of miles long. This halo is called the Corona, and is visible during total eclipses. By means of the spectroscope we know that there are about forty chemical elements in the sun, which are the same as those we find upon our earth.

As the sun weighs 330,000 times as much as the earth, the force of gravity upon its surface is twenty-seven and two-thirds times as much as it is here. A letter which weighs an ounce here would weigh almost a pound and three quarters on the sun; and a man of ordinary size in this world would weigh more than two tons there, and would be crushed to death by his own weight. Find how much your watch, your book, your pencil, your baseball, your football would weigh on the sun.

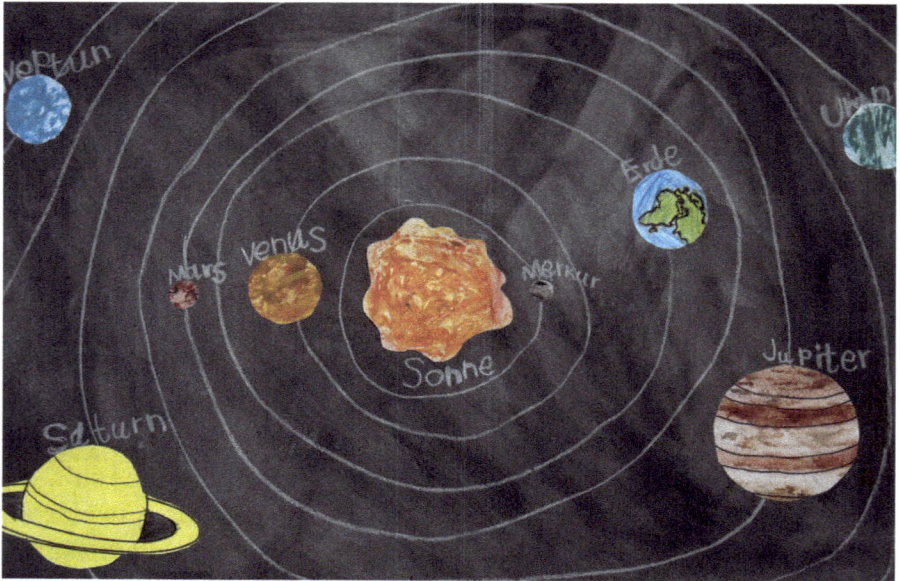

OUR SUN AND ITS FAMILY

First of all we shall have to acknowledge that our great, blazing sun is simply a medium-sized star, not nearly so large as Vega, nor even as large as the Pole-star; but it happens to be our own particular star and so is of the greatest importance to us. The sun has several other worlds, more or less like our own, revolving around it on almost the same level or plane in which our world revolves, but some of these worlds are much nearer the sun and others much farther away than ours. Nearest of all is Mercury, but it is not half so thick through as our earth, and it is so close to the sun that it circles around in 88 days; that is, its year is only 88 days long. Next comes Venus, almost as large as the earth, with a year 225 days long; next comes our earth, which completes its year in 365 days; next beyond us is Mars, a little more than half as thick as the earth and with a year 687 days long; beyond Mars is a group of small planets which are not large enough to be seen with the telescope, but we know that one of the largest of the group is only 490 miles through; beyond this mysterious swarm of little worlds is great Jupiter almost ten times as thick through as the earth, and it is so far away that it does not circle about the sun but once in 11 years; beyond great Jupiter comes Saturn, not quite ten times the diameter of the earth and so far from the sun that it takes 29½ years for it to move around its orbit; beyond Saturn is Uranus, only about four times as thick through as our world, and it has a year 84 years long; but the outermost of all our sun's planets is Neptune, little larger than Uranus, but so far from the sun that 165 years are required for it to complete its circle. Just think of a spring or a winter 41 years long! If Methuselah had lived on Neptune, he would have died before he was five and one-half years old.

Almost all of the Earth's sister planets are better off for moons than she; neither Venus nor Mercury has any moons. Mars has two moons, Jupiter five and Saturn has nine besides some splendid rings; and a queer thing about one of Saturn's moons is that it revolves in an opposite direction from the others. Uranus has four moons, while Neptune is not any better off than we are, unless, there are some we have not been able to discover because they are so far away.

One peculiar thing about all of the planets of the sun's family and all of their moons is that they all shine by reflecting the light of the sun, and none of them are hot enough to give off light independently; but these sister worlds of ours are so near us that they often seem larger and brighter than the stars, which are true suns and give off

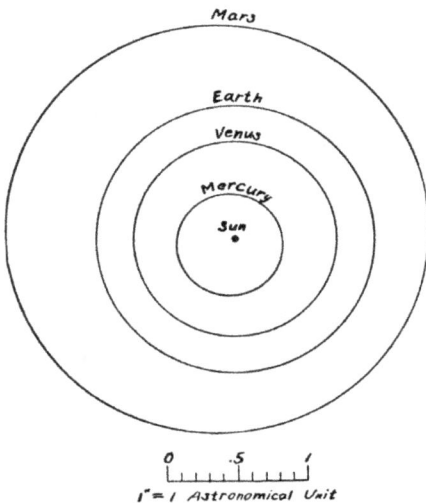

The orbits of the Inner Planets. Note that each planet has an orbit which is not circular but is very nearly so

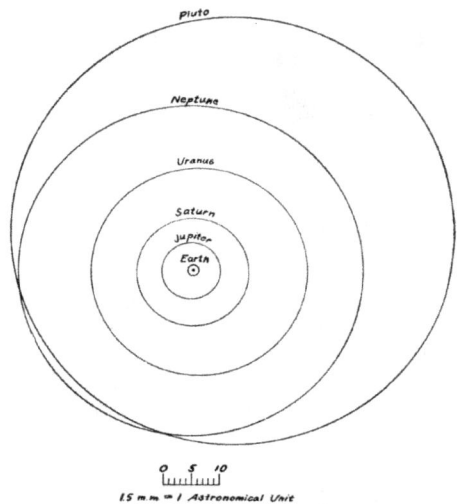

The Outer Planets and their orbits. The orbits of these planets also are not circular but are nearly so

much more light than our own sun. After a little experience the young astronomer learns to distinguish the planets from the true stars; the planets always follow closely the path of the sun and moon through the sky; they often seem larger and brighter than the true stars and do not twinkle so much. The so-called morning and evening stars are other planets of our sun's family and are not stars at all.

Dr. Simon Newcomb in his delightful book, "Astronomy for Everybody," gives the best illustration to make us understand the place of our sun and its planets and its relation to the stars in space. He explains that if here in the Atlantic States we should make a model of our solar system by putting an apple down in a field to represent the sun; then our earth could be represented by a mustard seed forty feet away revolving around the apple; and Neptune, our outermost planet, could be represented as a small pea circling around the apple at the distance of a quarter of a mile. Thus, our whole solar system could be modeled in a field one-half mile square, except for comets which might extend out in their long orbits for several miles. But to find the star nearest to our earth, the star that is only four and one-half light-years away from us, we should have to travel from this field across the whole of North America to California, and then take steamer and go out into the Pacific Ocean before we should reach our nearest star neighbor, which would be another sun like our own and be represented by another apple.

PHILIPP SALZGEBER (CC BY-SA 2.0)
The Hale-Bopp comet

COMETS

Besides planets and stars there are in space other bodies spinning around our great sun, and following paths shaped quite differently than those followed by our earth and its sister planets. We move around the sun nearly in a circle with the sun at the center, but these other heavenly bodies swing around in great ellipses, the sun being near one end of the ellipse and the other end being out in space beyond our farthest planet. These bodies do not revolve around the sun in the same plane as our world and the other planets, and indeed they often move in quite the opposite direction. The most noticeable of these bodies whose racetrack around the sun is long instead of circular are the comets, and we know that some of these almost brush the sun when turning at the end of their course. The astronomers have been able to measure the length of the race-tracks of some of the comets and thus tell when they will come back. Encke's comet, named after the German astronomer, makes its course in three and one-half years and this is the shortest period of any we know. There are about thirty comets whose courses have been thus measured; the longest period belongs to Halley's comet, which makes such a long trip that it comes back only once in seventy-six years; but there are other comets which astronomers are sure travel such long routes that they come back only once in hundreds or even thousands of years. About nine hundred comets have been discovered, many of them so small that they can only be seen through the aid of the telescope; and it has been found that in one instance, at least, three comets are racing around the sun on the same track.

A comet is a beautiful object, usually having a head which is a point of brilliant light and a long, flaring tail of fainter light, which always extends out from it on the side opposite the sun. The head of a comet must be nearly twice as thick through as the earth in order to be large enough for our telescopes to discover it. Some of the comet heads have been measured, and one was thirty-one times, and another one hundred and fifty times, as wide as our earth. If the heads are this large, imagine how long the tails must be! Some of them are far longer than the distance from our earth to the sun.

The head of a comet is supposed to be a mass of gas which is made to glow by the sun's heat, and is so volatile and thin that the heat evaporates it. In fact, this gas has so little weight that light can push it; one would never believe that light could push anything because we cannot feel it strike against us; but the physicists have found that it does push, and by pushing against the particles of the gas of comets it sends them out into a streamer away from the sun, just as the heat pushes out a flaring cloud of steam from the spout of a teakettle.

Another thing we know about comets is that they are not able to hold together, but break into pieces; and these pieces become cold out in space and condense and harden into lumps of metallic stone; and these lumps, each one whirling, follows the same track that the comet followed. If a comet should break into many pieces it would make a whole flock of these lumps all going in the same direction and in the same path about the sun.

Since comets are moving around the sun in every direction, it is possible that the earth may sometime meet one; and if this proves to be a "head on collision" there are those who prophesy that there will be no people left to tell the story; but the tails of comets are so thin and ethereal that our earth actually passed through one once, and no one but the astronomers knew anything about it.

SHOOTING STARS

When we look up during an evening walk and see a star falling through space, sometimes leaving a track of light behind it, we wonder which of the beautiful stars of the heavens has fallen. But astronomers tell us that no real star ever fell, but that what we saw was a lump of the matter of which worlds and comets are made; and it was following its own swift path around the sun, when by chance it crossed our earth's path, and was drawn toward us by that mysterious power called gravitation, which makes us fall down if we lose our balance, and which also made this bit of world-stuff fall to earth when it came so near us that its balance was disturbed. Although this shooting star was just a dark, cold lump of metal, too small for us to see, yet it was moving so swiftly along its path around the sun that the friction caused by its passing through our air, lighted it and burned it up, just as a match scratched on sandpaper lights and burns; as soon as it blazed

A meteor. The meteor, afterglow and wake can all be seen distinctly

we saw it and said, "There is a shooting star!" Sometimes the lump is so big that it does not have time to burn up while passing through the hundred miles or more of our atmosphere, and what is left of it strikes the earth usually with such force as to bury itself deep in the soil. Such lumps are called "meteoroids" before they fall and "meteors" while plunging white-hot through the air, but when they reach our earth we call what is left of them "meteorites." There are, in museums, many meteorites of this so-called stone, which is largely iron. Chemists find no new metals or elements in these strangers from space, but they do find new kinds of chemical partnerships and combinations. Some of these meteorites weigh hundreds of pounds, one in the Yale Museum weighing 1635 pounds. It surely would not be safe for a person to be on the spot where and when one of these meteorites strikes the earth; but there are so few of the meteors large enough to last until they become meteorites, that we may safely continue to enjoy the sight of shooting stars. If it were not for the air that wraps our globe, like a great, kindly blanket, and by its friction sets fire to the meteors and destroys them, no one could live on this earth because we all should be pelted to death. Prof. Newton estimated that every twenty-four hours our world meets seven millions of these shooting stars, some of them no larger than shot and others weighing tons.

The Relation between Comets and Meteors

It has been discovered that many of the shooting stars are gathered in great flocks and move about the sun in elongated paths, like the comets. We have learned the times of year when the path our earth follows comes close to these flocks of meteors which are flying around the sun like birds. One of these flocks is straggling, and we begin to meet it about the end of July and reach the center of the crowd on August 10th, and then continue to take stragglers until the last of August. We can see the point where we meet this flock of meteors, if we look for it in the direction of the constellation Perseus (see planisphere). On November 13th, we meet another flock which we find in the direction of the constellation Leo, of which the great star Regulus is the heart (see chart); but this flock is usually all in a bunch and we pass it in two days. Once there was a splendid flock which our world met every thirty-three years, and we took so many stragglers from it that our skies were filled with shooting stars, and ignorant people were greatly frightened; but for some reason, this flock has changed its path and we looked in vain for the great display of fireworks which was due to occur in 1899.

While we know from observation that the flocks of shooting stars, which make our star showers, are just broken pieces of comets which once traveled the same path, yet it does not follow that all our shooting stars are comet fragments. Prof. Elkins has shown by photographing meteors that some of them must be wanderers in the vast spaces which lie between the stars.

THE RELATION BETWEEN THE TROPIC OF CANCER AND THE PLANTING OF THE GARDEN[1]

BY JOHN W. SPENCER

A Story To Be Read to the Pupils

In years gone by, many farmers had a favorite phase of the moon when they planted certain crops, usually spoken of as the "dark" or the "light" of the moon. I once knew a woman who picked her geese by the "sign of the moon." Hogs were butchered in the "light" of the moon, and then the pork would not "fry away" so much in the skillet. It is true some pork from some hogs wastes faster than that of others, but the difference is due to the kind of food given the hogs. Many farmers hold to those old superstitions yet, but the number is much less now than twenty-five years ago. I wish I might impress on you young agriculturists that the moon has no influence on plant life, or pork, or geese, but the position of the sun most decidedly has. We have some plants that had best be planted when the sun's rays strike the state of New York slantingly, which means in early spring or late fall. We have other plants that should not be put in the open ground until the rays of the sun strike the state more direct blows, which means the hotter weather of summer. If I were in close touch with you pupils, I should be glad to tell some things that happen to three young friends of mine, hoping that thereby my statement might give the boys and girls an interest in three geographical lines concerning the tropics, and lead them to find their location on the map, particularly when later they learn what happens to my three young friends, whom we will call by the following names: There is one in Quito, Ecuador, of whom we will speak as Equator Shem; the one on the Island of Cuba is named Tropic of Cancer Ham; and the other in San Paulo, Brazil, answers to the name of Tropic of Capricorn Japhet.

What happens to these three boys, Shem, Ham and Japhet, is this. At certain times of the year they have no shadow when they go home for dinner at noon. This state of affairs is no fault of theirs. It is not because they are too thin to make shadows. It is due to the position of the sun. If the boys should look for that luminary at noon, they would find it as directly over their heads as a plumb line. It is a case of direct

---

1   A portion of a letter to apprentice gardeners from Uncle John, published as a supplement to the Home Nature-Study Course Leaflet, for April-May, 1907.

or straight blows from rays of the sun, and, oh, how hot—hotter than any Fourth of July the oldest inhabitant can remember! These three boys are not hit squarely on the head on one and the same day. Each is hit three months after the other. The first boy to be hit this year in the above manner will be the Equator Shem. The time will be during the last half of March. Can any of my young friends in this grade tell me the exact day of March that Equator Shem has no shadow? If no one of you can answer that question at this time, you had best talk it over with your friends, and bring your answers tomorrow. It happens at a time when our days are of about equal length.

Another thing about this particular day is that our almanacs call it the first day of spring. All because no boy or anything else has a shadow on the equator at noon time. People and bluebirds and robins in the state of New York will see squalls of snow about that time, and there will be some freezing nights. But after the first day of spring the cold storms do not last so long, as was the case during December, January, and early February, when the sun's rays hit us with very glancing blows. Watch to see how much faster the sun melts the snow on the last days of March than it did at Christmas time. The light is also stronger and brighter, and plants in greenhouses and our homes have more life, and are not so shiftless, so to speak. Even the hens feel the influence, for they begin to lay more eggs and cackle, and down goes the price of eggs. Do not forget to learn what day in March spring begins, when the Equator boy finds it so hot that he would like to take off his flesh, and sit in his bones. After a few days, Equator Shem will find he again has a shadow at noon. A short one it is true, but it will get longer and longer each day. Now his shadow will be on the south side of him. Is this a queer thing to happen? On which side of you is your noon-time shadow? I will give every one of you a red apple that finds it anywhere but on the north side of him at twelve o'clock. Every time the sun shines at noon, watch to find your old uncle in the wrong, and thereby get the apple. Each day that the shadow of Equator Shem becomes longer and longer, the noon-day shadow of Tropic of Cancer Ham, living on the Island of Cuba, will be getting shorter and shorter, until at last there comes a day during the last of June that he, too, will have no shadow, and the almanac says that that day is the beginning of summer.

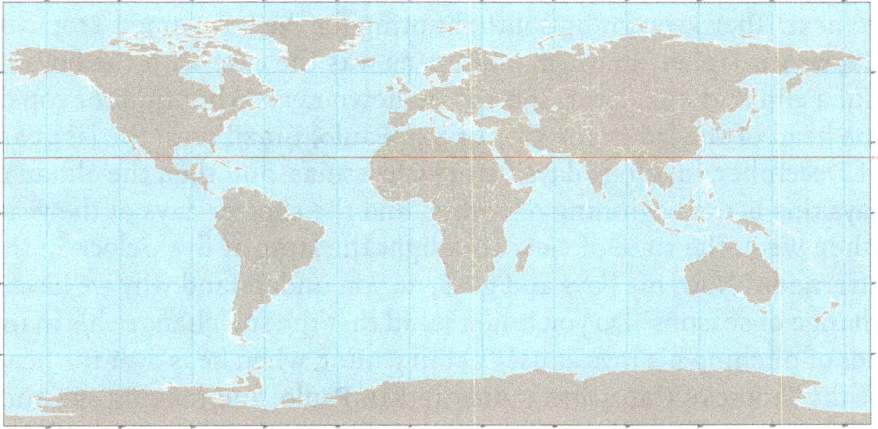

A map showing the tropic of cancer

Now it will be the turn of the Tropic of Cancer, Ham, on the Island of Cuba, to say the weather is hotter than two Fourths of July beat into one, and he too will wish that he could take off his flesh, and sit in his bones. Everybody in the state of New York will say that the first summer day is the longest day of the year. It is on this day that Equator Shem will have as long a shadow as *he* ever had in his life. No United States boy will ever be without a shadow at noon so long as he remains in his own country. When the eight o'clock curfew bell says it is time for boys and girls to go to bed, it will yet be light enough to read the papers. The sun not only sets late on that first summer day, but it appears early next morning. What a beautiful spectacle a sunrise in June is! Men of wealth will pay thousands of dollars for pictures showing its glory, yet I suppose that not one boy in five hundred ever saw the beauty of the birth of a new day in the sixth month of the year, and with no price of admission at that.

For only one day do the sun's rays fall directly on top of the head of Tropic of Cancer Ham, who lives on the Island of Cuba—just for one day, after which the up and down rays travel back towards the Equator Shem. On the twenty-first of September Shem again has no shadow at noon, and the almanac makers say that is the last day of summer, and tomorrow will be the first day of autumn. Again it is very hot where Shem lives, but the alligators and monkeys and the parrots do not seem to mind it. Where do the up and down rays of the sun go next? They keep going south, hunting for the boy named Tropic of Capricorn Japhet, to warm *him* up, as was the case with the boys in Cuba and at the Equator. The up and down rays do not find the top of the head of the lad in the City of San Paulo, Brazil until the last part of December, just four days before Christmas, and then the almanac says this is the beginning of winter, and the shorter days of the year, when we in the state of New York light the lamp at five o'clock in the afternoon. Now, my boys and girls, do you understand why we have a change of seasons? Do you understand that the sun changes his manner of pitching his rays at us? That in winter, when he is over the head of the Tropic of Capricorn Japhet in San Paulo, and making summer on that part of the earth, to us people in the north, in the State of New York, he pitches only slanting rays that do not hit us hard, and have but little power? Thus you will see that the rays of the sun that strike the earth direct blows, swing back and forth like a pendulum, year after year, and century after century, coming north as far as Tropic of Cancer Shem, but no farther, and then swinging south as far as the boy named Tropic of Capricorn Japhet, and no farther, just stopping and swinging back again towards the north.

The Zodiac and Its Signs

The mysterious symbols of the Zodiac on the first pages of almanacs are always a source of wonder and awe to children, and remain a life-long mystery to most people except fortune tellers; and yet the Zodiac is the simplest thing in the world to understand. However, the lesson should not be given until after the children have had their lessons on the sun and the shadow-stick, and also the lessons on the stars.

The ancients who believed the earth stood still and the sun moved around it, noticed inevitably that the path through the heavens pursued by the sun reached in summer a point farther north and higher up than in the winter, and they naturally wished to map this path, so as to fix it in their minds and writings. Nothing could be easier, for there in the skies were the eternal stars always following the same fixed path through the heavens and never wobbling up and down like the sun. So they chose the constellation which marked the highest point in the sun's path for each month, and these constellations might be likened to a stairway with six steps down toward the south and six steps up toward the north, the highest stair being reached by the sun in June, for then the sun is highest in the heavens and the farthest north. So beginning in June with Cancer, (the Crab), which is high in the heavens, it steps down to Leo, (the Lion) in July, takes another step lower to the Virgin in August, another down to the Scorpion in September, and comes to the lowest step of all, Sagittarius, (the Archer), in November; for at the last of November, the sun's path reaches its lowest point farthest south in the heavens and then the days are shortest. But in December it begins to climb and takes a short step up to Capricornus, (the Goat), in January it rises to Aquarius, (the Water Carrier), and in February rises another step to Pisces, (the Fishes). In March it reaches up to Aries, (the Ram), in April attains Taurus, (the Bull), and in May reaches Gemini, (the Twins), which step is almost as high and as near to the North Star as was the Cancer, where the journey began the June before.

It may be difficult for the pupils to learn to know all these constellations, as some of them are not very well marked; however, if they wish to learn them they can do so by the use of the planisphere.

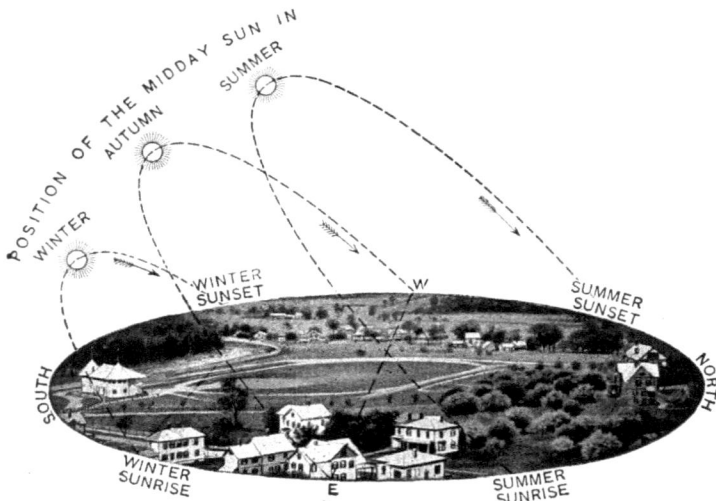

From Todd's New Astronomy

166

Some of the Zodiac constellations are marked by brilliant stars which have already been learned. Regulus is the heart of Leo, the Lion; Spica which means "ear" is the ear of wheat which the Virgin is holding in the constellation Virgo. Red Antares lies in the Scorpion; and the Milk Dipper, which is shaped like the Big Dipper, but smaller, marks Sagittarius. Red Aldebaran is the fiery eye of Taurus, the Bull, while the Gemini, or Twins, are the most conspicuous of the stars in the evening skies of February and March. It should be noted, however, that at the present day, owing to the peculiar movement of our earth, the path of the sun in climbing up and down these constellation steps is not quite the same as it seemed to the ancients.

A 6th century mosaic zodiac wheel in a synagogue

The Relations of the Sun to the Earth

TEACHER'S STORY
"Whether we look or whether we listen,
We hear life murmur or see it glisten."
—LOWELL.

ALL this murmuring and glistening life on our earth planet has its source in the great sun which swings through our skies daily, sending to us through the friendly ether his messages of light and warmth—messages that kindle life in the seed and perfect the existence of every living organism, whether it be the weed in the field or the king on his throne.

At sunrise this heat which the sun sends out equally at all times of day and night, is tempered when it reaches us because it passes obliquely through our atmosphere-blanket, and thus traverses a greater distance in the cooling air. The same is true at sunset; but at noon, when the sun is most directly over our heads its rays pass through the least possible distance of the atmosphere-blanket and, therefore, lose less heat on the way. It is true that often about three o'clock in the afternoon is the hottest period of the day, but this is because the air-blanket has become thoroughly heated; but we receive the most heat directly from the sun at noon.

The variations in the time of the rising and the setting of the sun may be made a most interesting investigation on the part of the pupils. They should keep a record for a month in the winter; and with this as a basis, use the almanac to complete the lesson. Thus, each one may learn for himself which is the shortest and which the longest day of the year. There is a slight variation in different years; the shortest day of the year when this lesson was written, as computed from a current almanac, was the 22d of December; it was nine hours and fourteen minutes long. The longest day of the year was the 22d of June, and it was fifteen hours and six minutes in duration. On the longest day of the year the sun reaches its farthest point north and is, therefore, most nearly above us at mid-day. On the shortest day of the year, the sun reaches its

Sunrise over the water

farthest point south and is, therefore, farther from the point directly above us at mid-day than during any other day of the year.

Also the movement of the sun north and south is an interesting subject for personal investigation, as suggested in the lesson. Through quite involuntary observation, I have become so accustomed to the arc traversed by the points of sunrise as seen from my home, that I can tell what month of the year it is, by simply noting the place where the sun rises. When it first peeps at us over a certain pine tree far to the south, it is December; when it rises over the reservoir it is February or October; and when it rises over Beebe pond it is July. Only at the equinox of spring and fall does it rise exactly in the east and set directly in the west. Equinox means equal nights, that is, the length of the night is equal to that of the day.

So vast is the weight of the sun that the force of gravity upon its surface is so great that even if it were not for the white-hot fireworks there so constantly active, we could not live upon it, for our own weight would crush us to death. But this multiplying the weight of common objects by twenty-seven and two-thirds to find how much they would weigh on the sun is an interesting diversion for the pupils, and incidentally teaches them how to weigh objects, and something about that mysterious force called gravity; and it is also an excellent lesson in fractions.

LESSON

Leading thought— The sun which is the source of all our light and heat and, therefore, of all life on our globe travels a path that is higher across the sky in June than the path which it follows in December, and hence we experience changes of seasons. The lesson should be given to the pupils of the upper grades and should be correlated with reading, arithmetic and thinking.

Observations—

1. What does the sun do for us?

2. At what time of the day after the sun rises do we get the least heat from it? What hour of the day do we get the most heat from it?

3. Is the sun equally hot all day? Why does it seem hotter to us at one time of the day than at another?

4. At what hour does the sun rise and set on the first of the following months; February, March, April, May and June?

5. Which is the shortest day of the year, and how long is it?

6. Which is the longest day of the year, and how many hours and minutes are there in it?

Earth at the March 2019 equinox

7. What day of the year is the sun nearest a point directly over our heads at mid-day?

8. Which day of the year is the sun at mid-day farthest from the point directly above our heads? Explain why this is so.

9. Standing in a certain place, mark by some building, tree or other object just where the sun

rises in the east and sets in the west on the first of February. Observe the rising and setting of the sun from the same place on the first day of March and again on the first of April. Does it rise and set in the same place always or does it move northward or southward?

10. Is the sun farthest south on the shortest day of the year? If so, is it farthest north on the longest day of the year?

11. At what time of the year does the sun rise due east and set due west?

12. The sun is so much larger than the earth that its force of gravity is twenty-seven and two-thirds times that of the earth. How much would your watch weigh if you were living on the sun? How much would you yourself weigh if you were there?

13. *Experiment. A shadow stick*— Place a peg two or three inches high upright in a board and place the board lengthwise on the sill of a south window or where it will get the south light. Note the length cast by the shadow of the peg during a sunny day and draw a line with pencil or chalk outlining the tip of the shadow of the stick from 9 a.m. to 4 p.m. Make a similar outline a month later and again a month later and note whether the shadow traces the same line during each of these days of observation. Note especially the length of the shadow at noon.

Another excellent observation lesson for teaching the fact that the sun travels farther south in the winter, is to measure the shadow of a tree on the school grounds at noonday once a month during the school year. The length of the tree shadow can be measured from the base of the tree trunk, a memorandum being made of it.

14. When does the stick or tree cast its longest shadow at noon—in December or February? February or April? April or June? Why?

Topics for English themes— The size and distance of the sun. The heat of the sun and its effect upon the earth. What we know about the sun spots. Our path around the sun.

Supplementary reading—
Starland, Ball; *The Earth and Sky*, Holden.

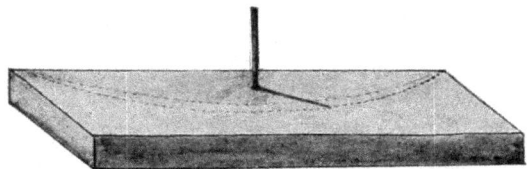

A shadow-stick.

How To Make a Sundial

Method— The diagram for the dial is a lesson in mechanical drawing. Each pupil should construct a gnomon *(no-mon)* of cardboard, and should make a drawing of the face of the dial upon paper. Then the sundial may be constructed by the help of the more skillful in the class. It should be made and set up by the pupils. A sundial in the school grounds may be made a center of interest and an object of beauty as well.

Materials— For the gnomon a piece of board a half inch thick and six inches square is required. It should be given several coats of white paint so that it will not warp. For the dial, take a board about 14 inches square and an inch or more thick. The lower edge may be bevelled if desired. This should be given three coats of white paint, so that it will not warp and check.

To make the gnomon— The word gnomon is from a Greek word meaning "one who knows." It is the hand of the sundial, which throws its shadow on the face of the dial, indicating the hour. Take a piece of board six inches square, and be very sure its angles are right angles.

172

Let s, t, u, v represent the four angles; draw on it a quarter of a circle from s to u with a radius equal to the line vs. Then with a cardboard protractor, costing fifteen cents, or by working it out without any help except knowing that a right angle is 90°, draw the line vw making the angle at x the same as the degree of latitude where the sundial is to be placed. At Ithaca the latitude is 42° 27' and the angle at x measures 42° 27'.

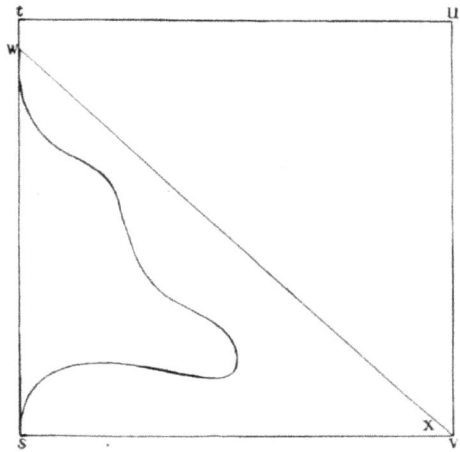

The gnomon.

Then the board should be cut off at the line vw, and later the edge sw may be cut in some ornamental pattern.

To make the dial— Take the painted board 14 inches square and find its exact center, y. Draw on it with a pencil the line A A" a foot long and one-fourth inch at the left of the center. Then draw the line B B" exactly parallel to the line A A" and one-fourth inch to the right of it. These lines should be one-half inch apart—which is just the thickness of the gnomon. If the gnomon were only one-fourth inch thick, then these lines should be one-fourth inch apart, etc.

With a compass, or a pencil fastened to a string, draw the half-circle A A' A" with a radius of six inches with the point C for its center. Draw a similar half-circle B B' B" opposite with C' for its center. Then draw the half-circle from D, D', D", from c with a radius of five and three-quarter inches. Then draw similarly from c' the half-circle E, E', E". Then draw from c the half-circle F, F', F" with a radius of five inches and a similar half-circle G, G', G" from c' as a center.

Find the points M, M' just six inches from the points F, G; draw the line J, K through M, M' exactly at right angles to the line A, A'. This will mark the six o'clock point so the figures VI may be placed on it in the space between the two inner circles. The noon mark XII should be placed as indicated (the "X" at D, F, the "II" at E, G). With black paint outline all the semi-circles and figures.

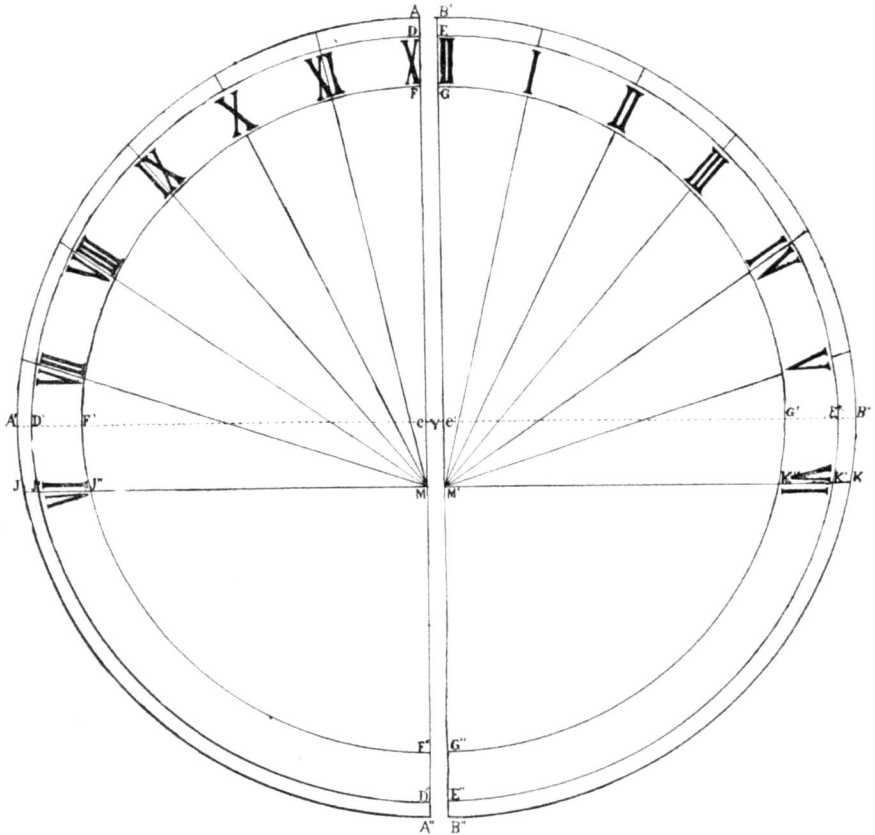

The face of the sundial.

To set up the sundial— Fasten the base of the gnomon by screws or brads to the dial with the point s of the gnomon at F, G, and the point v of the gnomon at M, M', so that the point W is up in the air. Set the dial on some perfectly level standard with the line A, A" extending exactly north and south. If no compass is available, wait until noon and set the dial so that the shadow from W will fall exactly between the points A, B, and this will mean that the dial is set exactly right. Then with a good watch note the points on the arc E, K', on which the shadow falls at one, two, three, four, and five o'clock: and in the morning the points on the arc J' D on which the shadow falls at seven, eight, nine, ten and eleven o'clock. Draw lines from M to these points, and lines from M' to the points on the arc E K'. Then place the figures on the dial as indicated in the spaces between the two inner circles. The space between the two outer circles may be marked with lines indicating the half and quarter hours. The figures should be outlined in pencil and then painted with black paint, or carved in the wood and then painted.

Twilight, twilight of the west,
Sky-lines fading into rest,
Cloud-bars lying far and slight,
Shadows sinking into night,—
O moon, ye moon, so faint and still,
Hanging, hanging as ye will
Low along the western sky,
Far and far and yet so nigh
A finger's breadth within the sheen
And silent shoreless vasts between—
Thy aching heart is long ages lost,
And clear and calm as film of frost,
Ye know no longer strain or stress,
All passionless and passionless.
—FROM "THE NEW MOON," L. H. BAILEY.

Full moon in the darkness of the night sky

The Moon

TEACHER'S STORY

THE moon is in more senses than one an illuminating object for both the earth and the skies. As a beginning for earth study it is an object lesson, illustrating what air and water do for our world and incidentally for us; while as the beginning of the study of astronomy, it is the largest and brightest object seen in the sky at night; and since it lies nearest us, it is the first natural step from our world to outer space.

The moon is a little dead world that circles around our earth with one face always towards us, just as a hat-pin thrust into an apple would keep the same side of its head always toward the apple no matter how rapidly the apple was twirled. As we study the face of the moon, thus always turned toward us, we see that it is dark in some places and shining in others, and some ignorant people have thought that the dark places are oceans and the light places, land. But the dark portions are simply areas of darker rocks, while the lighter portions are yellowish or whitish rocks. The dark portions are of such a form that people have imagined them to represent the eyes, nose and mouth of a man's face; but a far prettier picture is that of a woman's uplifted face in profile. The author has a personal feeling on this point, for as a child she saw the man's face always and thought it very ugly and, moreover, concluded that he chewed tobacco; but after she had been taught to find the face of the lady, the moon was always a beautiful object to her.

The moon is a member of our sun's family, his grand-daughter we might call her if the earth be his daughter; and since the moon has no fires or light of its own, it shines by light reflected from the sun and, therefore, one-half of it is always in shadow. When we see the whole surface of the lighted half we say the moon is full; but when we see only half of the lighted side turned toward us,

The lady in the moon

we say the moon is in its quarter, because all we can see is one-half of one-half which is one-quarter; and when the lighted side is almost entirely turned away from us we say it is a crescent moon; and when the lighted side is entirely turned away from us we say there is no moon, although it is always there just the same. Thus, we can understand that, although we can never see the other side of the moon, the sun shines on all sides of it. Our earth, like the moon, shines always by reflected light and is almost four times as wide as the moon. Think what a splendid moon our earth must seem to the lady in the moon! When we see the old moon in the new moon's arms, the dark outline of the moon within the bright crescent is visible because of the earth-shine reflected from it. Sometimes pupils confuse this appearance of the moon with a partial eclipse; but the former is the new or old moon, which is one edge of the moon shining in the sunlight, the remainder faintly illumined by earth light, while an eclipse must always occur at the full of the moon when the earth passes between the sun and the moon, hiding the latter in its shadow.

It is approximately a month from one new moon to the next, since it takes twenty-nine and one-half days for the moon to complete its cycle around the earth and thus turn once around in the sunshine. Therefore, each moon day is fourteen and three-quarter days long and the night is the same length. The moon always rises in the east and sets in the west, following pretty nearly the sun's summer path. The full moon rises at sunset and sets at sunrise, but owing to the movement of the earth around the sun the moon rises about fifty minutes later each evening; however, this time varies with the different phases of the moon and at different times of the year. This difference in the time of rising is so shortened in August, that we have several nights when the full moon lengthens the day; and it is called the "harvest moon," because in England it adds to the hours devoted to harvesting the grain.

A Visit to the Moon

If we could be shot out from a Jules Verne cannon and make a visit to the moon, it would be a strange experience. First, we should find on this little world, which is only as thick through as the distance from Boston to Salt Lake City, mountains rising from its surface more than thirty thousand feet high, which is twice as high as Mt. Blanc and a thousand feet higher than the tallest peak of the Himalayas; and these moon mountains are so steep that no one could climb them. Besides ranges of these tremendous mountains, there are great craters or circular spaces enclosed with steep rock walls many thousand feet high. Sometimes at the center of the crater there is a peak lifting itself up thousands of feet, and sometimes the space within the crater circle is level. Thirty-three thousand of these craters have been discovered.

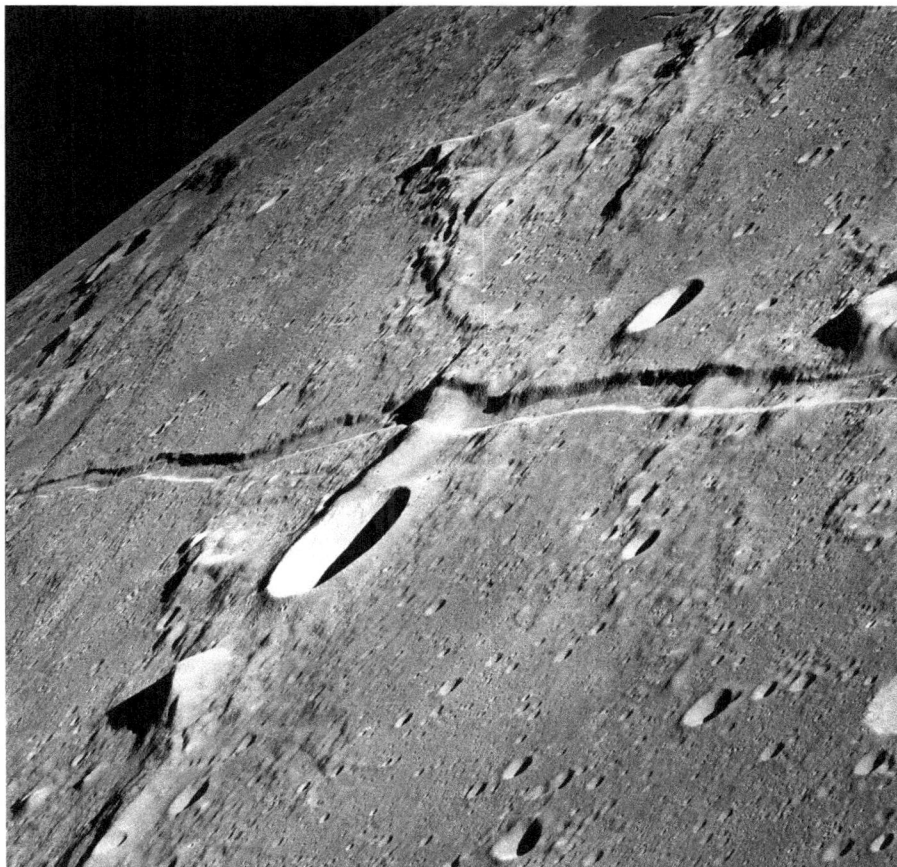

The surface of the moon

And, too, on the moon, there are great plains and chasms; and all these features of the moon have been mapped, measured and photographed by people on our earth. For a boy studying geometry, the measuring of the height of the mountains of the moon is an interesting story.

But we could never in our present bodies visit the moon, because of one terrible fact—the moon has no air surrounding it. No air! What does that mean to a world? First of all, as we know life, no living thing—animal or plant—could exist there, for living beings must have air. Neither is there water on the moon; for if there were water there would have to be air. And without water no green thing can be grown, and the surface of the moon is simply naked, barren rock.

If we were on the moon, we could not turn our eyes toward the

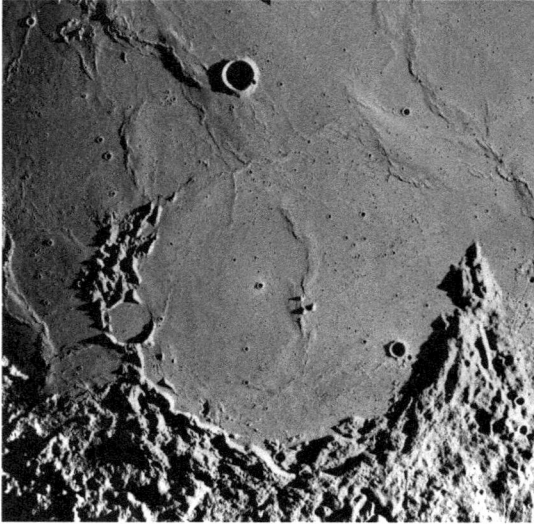
Wrinkle ridges in a moon crater

sun, for with no air to veil it, its fierce light would blind us; and the sky is as black at midday as at midnight, since there is no atmosphere to sift out the other rays of light, leaving the beautiful blue in the sky; nor is there a glow at sunset because there is no air prism to separate the rays of light and no clouds to reflect or refract them. The stars could be seen in the black skies of midday as well as in the black skies of night, and they would be simply points of light and could not twinkle, since there is no air to diffuse the sun's light and thus curtain the stars by day and cause them to twinkle at night. The shadows on the moon are, for the same reason, as black as midnight and as sharply defined; and if we should step into the shadow of a rock at midday we should be hidden as much as if we had stepped into a well of ink, or put on the invisible cloak of fairy lore. And because of no layers of air to make an aerial perspective, a mountain a hundred miles away would seem as close to us as one a mile away.

Since there is no atmosphere on the moon to act as a buffer between the cold of outer space, which is estimated to be 250° below zero, and the heat of the sun, which is 500° above zero, the temperature of the moon would vary 750° between day and night, or between sunshine and shadow, because there is no air to carry the heat over into the shadow or to blanket the world at night. But this great change of temperature between sunlight and darkness is the only force on the moon to change the shape of its rocks, for the expansion under heat and contraction under cold must break and crumble even the firmest rock more or less. Our rocks are broken by the freezing of water that creeps into every crevice, but there is no water to act on the moon's mountains in this fashion or to wear them away by dashing over their surface. However, the rocks and mountains of the moon may be changed in shape by the battering of meteorites, which pelt into the moon by the million, since the moon has no air to set them afire and make them into harmless shooting stars, burning up before they strike. But though a meteorite weighing thousands of tons should crash into a moon mountain and shatter it to atoms there would be no sound, since sound is carried only by the atmosphere.

Imagine this barren, dead world, chained to our earth by links forged from unbreakable gravity, with never a breath of air, a drop of rain or flake of snow, with no streams, nor seas, nor graced by any green thing—not even a blade of grass—a tree, nor by the presence of any living creature! Out there in space it whirls its dreary round, with its stupendous mountains cutting the black skies with their jagged peaks above, and casting their inky shadows below; heated by the sun's rays until hotter than the flame of a blast furnace, then suddenly immersed into cold that would freeze our air into solid ice, its only companion the terrific rain of meteoric stones driven against it with a force far beyond that of cannon balls, and yet with never a sound as loud as a whisper to break the terrible stillness which envelops it.

LESSON

Leading thought— The moon always has the same side turned toward us so we do not know what is on the other side. The moon shines by reflected light from the sun, and is always half in light and half in shadow. The moon has neither air nor water on its surface and what we call the moon phases depend on how much of the lighted surface we see.

Method— Have the pupils observe the moon as often as possible for a month, beginning with the full moon. After the suggested experiment, the questions which follow may be given a few at a time.

Experiment for recess— Darken the room as much as possible; use a lighted lamp or gas jet or electric light for the sun, which is, of course, stationary. Take a large apple to represent the earth and a small one to represent the moon. Thrust a hat pin

181

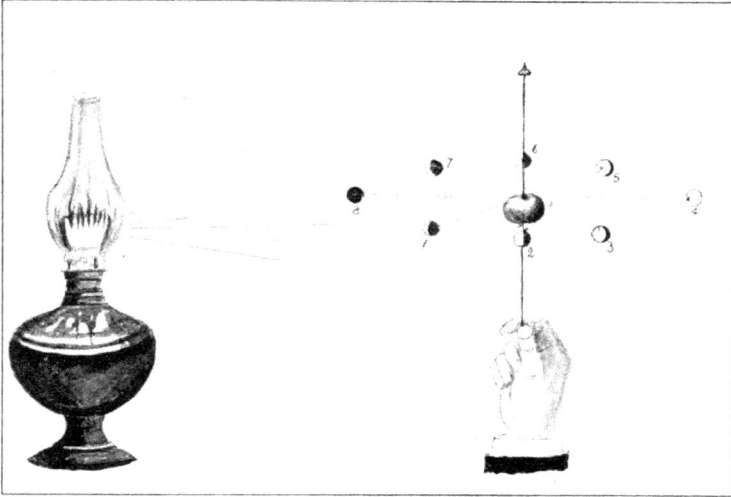
Experiment for illustrating the phases of the moon

through the big apple to represent the axis of the earth and also the axis about which the moon revolves. Tie a string about a foot long to the stem of the moon apple and make fast the other end to the hat pin just above the earth apple. Hold the hat pin in one hand and revolve the apple representing the moon slowly with the other hand letting the children see that if they were living on the earth apple the following things would be true:

1. Moving from right to left when the moon is between the earth and the sun it reflects no light.

2. Moving a little to the left a crescent appears.

3. Moving a quarter around shows the first quarter.

4. When just opposite the lamp, it shows its whole face lighted turned toward the earth.

5. Another quarter around shows a half disc, which is the third quarter.

6. When almost between the sun and the earth the crescent of the old moon appears.

7. Note that the moon always keeps one face toward the earth.

8. Note that the new moon crescent is the lighted edge of one side of the moon, while the old moon crescent is the lighted edge of the opposite side.

9. Make an eclipse of the moon by letting the shadow of the earth

A crescent and a half moon

fall upon it, and an eclipse of the sun by revolving the moon apple between the sun and the earth. The earth's orbit and the moon's orbit are such that this relative position of the two bodies occurs but seldom.

Observations—

1. Describe how the moon looks when it is full.

2. What do you think you see in the moon?

3. Describe the difference in appearance between the new moon and the full moon, and explain this difference.

4. Where does the new moon rise and where does it set?

5. When does it rise and when does it set?

6. Where and when does the full moon rise and where and when does it set?

7. How does the old moon look?

8. Could the crescent moon which is seen in early evening be the old moon instead of the new; and, if not, why not?

9. When and where do we ordinarily see the old moon when it is crescent shaped?

10. Does the moon rise earlier or later on succeeding nights? What is approximately the difference in time of moonrise on two successive nights?

11. Do you think we always look at the same side of the moon? If so, why?

12. Is more than one side of the moon luminous? Why?

13. How many days from one new moon until the next?

14. How long is the day on the moon and how long the night?

15. How many times does the moon go around the earth in a year?

16. What is the difference between the disappearance of the old moon and an eclipse of the moon? In both cases the moon is hidden from us.

The Physical Geography of the Moon

Questions for the pupils to think about and answer if they can—

17. Since it has been proved that there is no air or water on the moon, could there be any life there?

18. Supposing you could do without air or water and should be able to visit the moon, what would you find to be the color of the sky there?

19. Would there be a red glow before sunrise or beautiful colors at sunset?

20. Would the sun appear to have rays? Could you look at the sun without being blinded?

21. Would the stars appear to twinkle? Could you see the stars in the daytime?

22. How would the shadows look? If you could step into the shadow of a rock at midday, could you be seen?

23. Could you tell by looking at it whether a mountain was far or near?

24. It is estimated that the temperature of outer space is 250 degrees below zero, and the sun's direct heat is 500 degrees above zero. If this be correct, how hot would it be in the sunshine on the moon? How cold would it be at midnight?

25. Why is it so much hotter and colder on the moon than upon the earth?

26. If you could shout on the moon, how would it sound? If one hundred cannons should be fired at once on the moon, how would it sound?

27. Is there any rain or snow on the moon? Are there any clouds there? If there are no air and water on the moon, would the intense

heat and the powerful cold affect the soils or rocks, as freezing and thawing affect our rocks?

28. Professor Newton estimated that the earth meets seven million meteorites (shooting stars) every twenty-four hours. Why do we not see more of these? What happens when a meteorite strikes the moon?

29. The moon is so small that the force of gravity on its surface is one-sixth that on the earth's surface. If a man can carry seventy-five pounds on his back here, how much could he carry on the moon? If a boy can throw a ball one hundred yards here, how many yards could he throw on the moon? If a boy can kick a football one hundred and thirty-five feet in the air here, how far could he kick it on the moon?

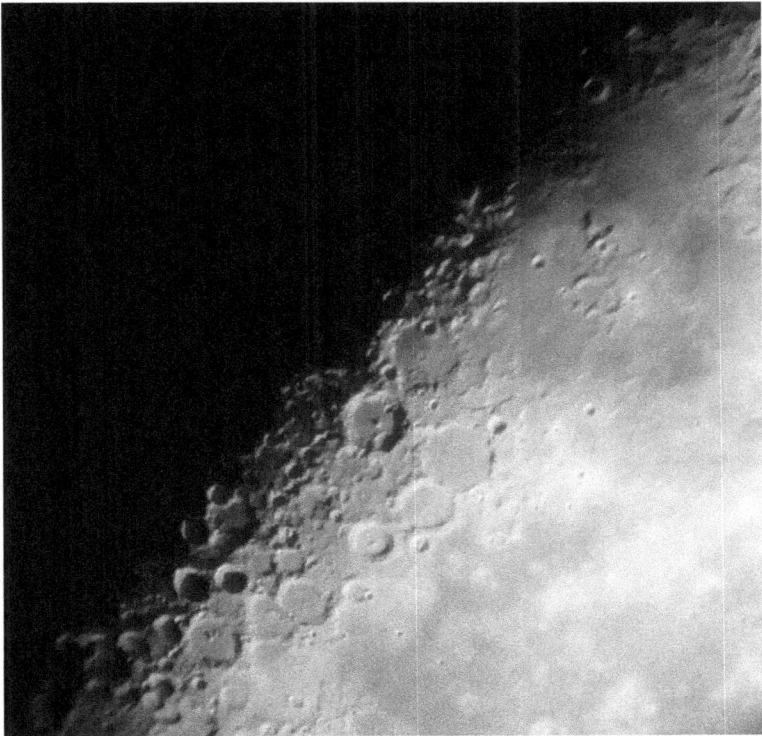

ÆSTRONOMÆR (CC BY-SA 4.0)

Lunar craters

www.ingramcontent.com/pod-product-compliance
Lightning Source LLC
Chambersburg PA
CBHW060224030426
42335CB00014B/1335